AF323575

SINGLE VARIABLE
CalcLabs
WITH THE TI-85/86

for Stewart's

FOURTH EDITION

CALCULUS
SINGLE VARIABLE CALCULUS
CALCULUS: EARLY TRANSCENDENTALS
SINGLE VARIABLE CALCULUS: EARLY TRANSCENDENTALS

Jeff Morgan
Texas A & M University

David Rollins
University of Central Florida, Orlando

BROOKS/COLE PUBLISHING COMPANY

I(T)P® An International Thomson Publishing Company

Pacific Grove • Albany • Belmont • Bonn • Boston • Cincinnati • Detroit • Johannesburg • London
Madrid • Melbourne • Mexico City • New York • Paris • Singapore • Tokyo • Toronto • Washington

A GARY W. OSTEDT BOOK

Assistant Editor: *Carol Ann Benedict*	Production Coordinator: *Dorothy Bell*
Marketing Manager: *Caroline Croley*	Cover Illustration: *dan clegg*
Marketing Assistant: *Debra Johnston*	Printing and Binding: *West Publishing*

COPYRIGHT © 1999 by Brooks/Cole Publishing Company
A division of International Thomson Publishing Inc.
I(T)P   The ITP logo is a registered trademark used herein under license.

For more information, contact:

BROOKS/COLE PUBLISHING COMPANY
511 Forest Lodge Road
Pacific Grove, CA 93950
USA

International Thomson Editores
Seneca 53
Col. Polanco
11560 México, D. F., México

International Thomson Publishing Europe
Berkshire House 168-173
High Holborn
London WC1V 7AA
England

International Thomson Publishing GmbH
Königswinterer Strasse 418
53227 Bonn
Germany

Thomas Nelson Australia
102 Dodds Street
South Melbourne, 3205
Victoria, Australia

International Thomson Publishing Asia
60 Albert Street
#15-01 Albert Complex
Singapore 189969

Nelson Canada
1120 Birchmount Road
Scarborough, Ontario
Canada M1K 5G4

International Thomson Publishing Japan
Palaceside Building, 5F
1-1-1 Hitotsubashi
Chiyoda-ku, Tokyo 100-0003
Japan

All rights reserved. No part of this work may be reproduced, stored in a retrieval system, or
transcribed, in any form or by any means—electronic, mechanical, photocopying, recording, or
otherwise—without the prior written permission of the publisher, Brooks/Cole Publishing
Company, Pacific Grove, California 93950. You can request permission to use material from this
text through the following phone and fax numbers:
Phone: 1-800-730-2214   Fax: 1-800-730-2215

Printed in the United States of America

10 9 8 7 6 5 4 3 2 1

ISBN 0-534-36436-5

Contents

Introduction

The TI-85 and TI-86 graphing calculators are powerful tools which can help students tackle many of the problems encountered in a typical single variable calculus course. These calculators have amazing plotting and solving capabilities that can free students from tedious calculations, and allow them to focus on the concepts of calculus.

This manual was created as a supplement for the fourth edition of James Stewart's **Calculus**. It is not intended to be a calculus text. It is simply a collection of examples, exercises and projects which can help students and instructors blend the use of the TI-85/86 calculators into the calculus curriculum. Chapters 1-8 introduce many of the commands and features of the TI-85/86 calculators. These chapters contain instruction and exercises associated with the basic calculus-related commands which are available on the TI-85/86. Chapter 9 gives a brief introduction to basic programming and program execution. Finally, Chapter 10 is a collection of projects which cover the spectrum between application and theory. These projects vary in difficulty, and require students to combine the use of hand calculation with technology.

NOTES FOR TI-85 USERS: Although this manual can be used with either the TI-85 or the TI-86, the command sequences throughout the manual refer to the newer model TI-86. The TI-85 can perform all of the calculations in this manual except the plotting of direction fields for differential equations. TI-85 users simply need to be aware that some of the commands are on different function keys than they are on the TI-86. For example, in the **MATH** submenu of the graphing screen, the **ROOT** command is $\boxed{\textbf{F1}}$ on the TI-86 and $\boxed{\textbf{F3}}$ on the TI-85. So, if the function key described in the text doesn't match the TI-85, look carefully at the other function keys, or use $\boxed{\textbf{MORE}}$ to move around the menu to locate it. Also note that the word **RANGE** is used in place of **WIND** for setting the window size of the

GRAPH screen for the TI-85. In addition, there are some keys that are at different locations on the TI-85. For example, **CATALOG** and **VARS** are on separate keys on the TI-85, and the tolerance settings have a separate key. If you are having trouble finding a particular key or command as described in the text, check your user's manual for the correct keystrokes.

There are also some differences in the calculator/user interface. For example, many of the commands from the **GRAPH** screen require setting a left and right bound to define an interval. Intervals are needed for calculating, among other things, intersections, critical points and inflection points. For the TI-86, the user is prompted for the bounds, but with the TI-85, these are set by the variables **LOWER** and **UPPER**. To set these, enter the **MATH** submenu and press either $\boxed{\textbf{F1}}$ for **LOWER** or $\boxed{\textbf{F2}}$ for **UPPER**. Use the arrow keys to locate the bound and press $\boxed{\textbf{ENTER}}$.

Comments and/or corrections should be sent to Jeff Morgan, Department of Mathematics, Texas A&M University, jmorgan@math.tamu.edu.

Chapter 1

Getting Started

This chapter introduces some of the features of the TI-86 that will be important throughout the remainder of this manual. This chapter is not intended to be a substitute for the **TI-86 Guidebook**. Always keep your **TI-86 Guidebook** handy for reference. Moreover, you should work carefully through Chapter 1 and make at least a cursory pass through Chapters 2 and 3 of your **TI-86 Guidebook**. In addition, read Chapter 1 in Stewart's **Calculus** prior to reading section 1.3 below.

1.1 Basic Calculations

The home screen is the starting point for most numerical computations on the TI-86. Let's clear the home screen (press $\boxed{\textbf{CLEAR}}$) and work a few simple examples.

Basic arithmetic can be performed on the TI-86. For example, try calculating

$$\frac{16.73 - 13.2\,(21.3 - \pi/12)}{5.7 + \pi}$$

The screen on the left below shows the expression before pressing $\boxed{\textbf{ENTER}}$; the screen on the right shows the expression after pressing $\boxed{\textbf{ENTER}}$. (Note that π is typed by pressing $\boxed{\textbf{2nd}}\ \boxed{\text{\^{}}}$ and $\sqrt{\ }$ is typed by pressing $\boxed{\textbf{2nd}}\ \boxed{x^2}$.)

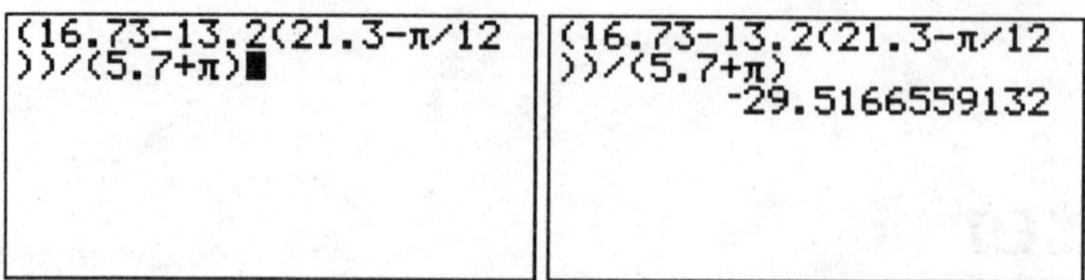

Always look closely at the input line. If an expression is entered incorrectly, it can be modified by using a combination of the arrow keys, $\boxed{\textbf{DEL}}$, $\boxed{\textbf{2nd}}$ [**INS**] and $\boxed{\textbf{CLEAR}}$ (to wipe out the entire expression).

It is very important to be careful with parentheses. The TI-86 will not read your mind. You must be precise. It is also important to use the *proper type* of parentheses. Different parentheses have different uses on the TI-86. Standard calculations use (*round*) parentheses, vector and matrix calculations use [square] parentheses, and list calculations use {*set*} parentheses.

Calculations involving the basic mathematical functions can also be done on the TI-86. In fact, the TI-86 knows almost every standard mathematical function. However, before any calculations involving trigonometric functions are performed, you should check whether the calculator will treat arguments as radians or degrees. Most trigonometric computations in Calculus should be done in radians. To specify this on the TI-86, press $\boxed{\textbf{2nd}}$ [**MODE**] and select **Radian** as shown below. Use the cursor keys to highlight **Radian**, press $\boxed{\textbf{ENTER}}$, and finally press $\boxed{\textbf{EXIT}}$ to return to the home screen.

For example, let's compute $\sin(1.32) + 2\cos(\pi/12)$ in **Radian** mode.

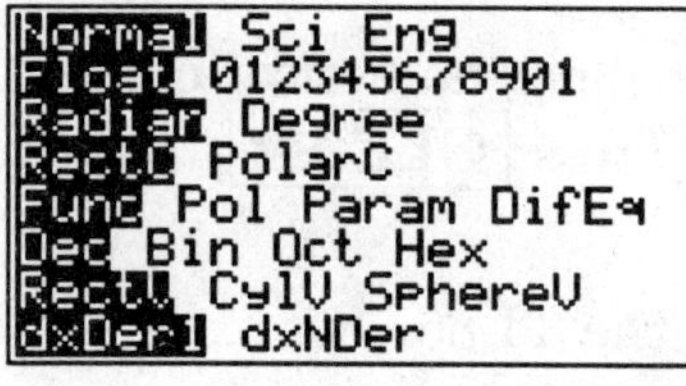

Notice that the result is much different if the calculator is in **Degree** mode.

Now, reset your calculator to **Radian** mode.

There are a number of special TI-86 commands which are used very frequently. Three of the most useful commands are ▶ **Frac**, 2nd [ENTRY] and 2nd [ANS]. The ▶ **Frac** command is used to convert the output of a calculation to a fraction. For example, suppose you want to calculate

$$\frac{234 - 321}{52} + 16$$

and have the result in the form of a fraction. Simply perform the calculation and then access the ▶ **Frac** command from the **MISC** menu on the **MATH** screen by pressing 2nd [MATH] F5 MORE F1.

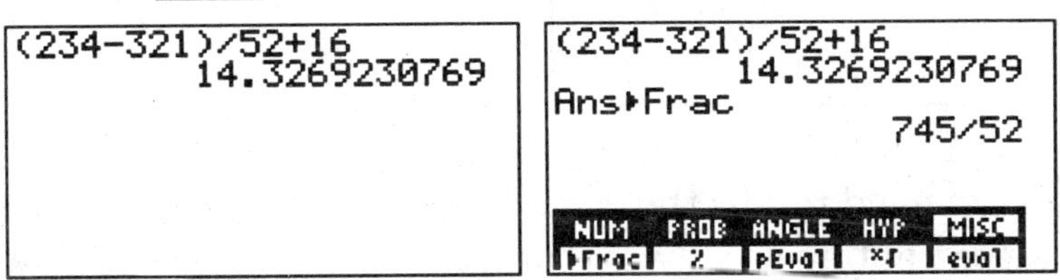

Let's look more carefully at the process above. Pressing 2nd [MATH] gives the screen below.

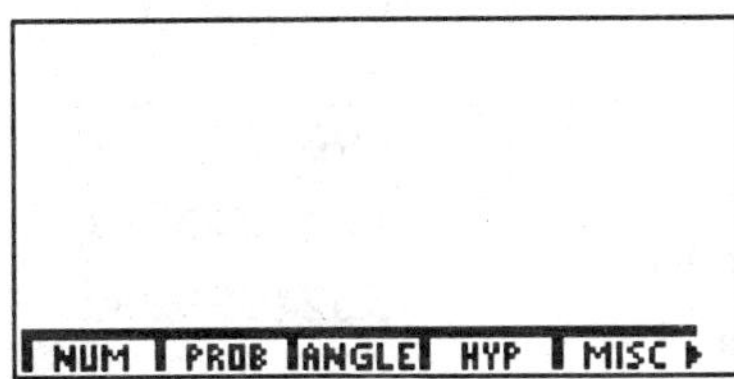

The menus which appear across the bottom of the screen can be accessed using the F1 through F5 keys. Consequently, pressing F5 accesses the **MISC** menu. Notice that two layers of menus are created in this process.

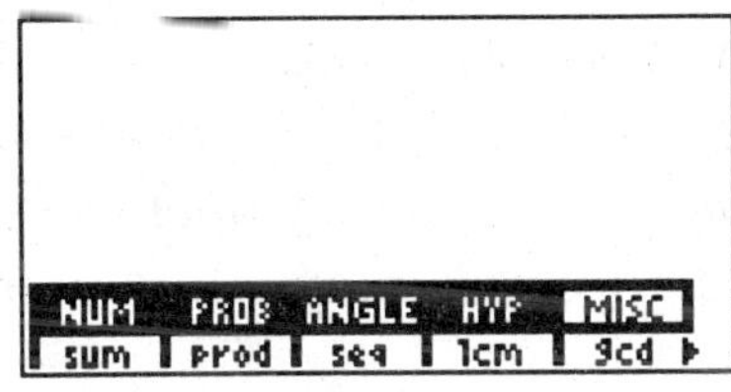

The top layer of menus can be accessed by pressing $\boxed{\text{2nd}}$ [M1] through $\boxed{\text{2nd}}$ [M5]. Press $\boxed{\text{EXIT}}$ twice (once for each layer of menu commands) to remove these menus from the home screen.

NOTE: The character ▶ appears at the end of the menus in each of the screens above. This indicates that other menus are available. These can be accessed by pressing $\boxed{\text{MORE}}$.

The **ENTRY** command can be used to make use of (up to 10) previous input lines. Press $\boxed{\text{2nd}}$ [ENTRY] to toggle the previous input lines to the screen. These lines can be re-executed by pressing $\boxed{\text{ENTER}}$, or modified and executed. The two screens below show the result of pressing $\boxed{\text{2nd}}$ [ENTRY] twice.

```
Ans▶Frac
```

```
(234-321)/52+16■
```

The **Ans** command recalls the output from the previous calculation. **Ans** is sometimes used automatically by the calculator. We saw this above with the ▶ **Frac** command. This command can also be accessed for direct calculation. For example, if we compute $\sin(\pi/12) + \cos(\pi/12) + \sqrt{2}$ and decide to calculate the square root of the result, then we can press $\boxed{\text{2nd}}$ [ANS] as shown below.

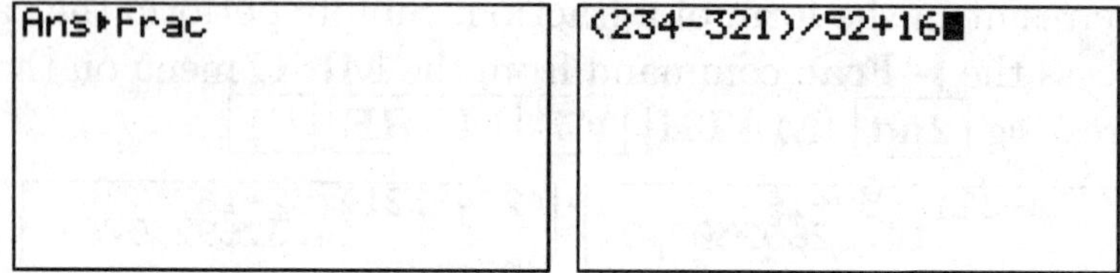

Notice that the first input line in the screen above has the exact appearance as the expression we are trying to compute. The TI-86 has been designed so that *almost* any expression can be entered exactly as it is written on a piece of paper. We say "almost", since, we must be careful of the order in which the calculator evaluates different operations. The list below gives the most important ordering for this manual. A more detailed list is given in the **TI-86 Guidebook**.

- Function keys, $\boxed{x^2}$ and $\boxed{\textbf{2nd}}$ $[x^{-1}]$ which are pressed after you enter the value of x.

- Powers and roots, such as $\boxed{\textbf{2nd}}$ $[e^x]$ and the exponentiation key $\boxed{\char94}$.

- Function keys like $\boxed{\textbf{2nd}}$ $[\sqrt{\ }]$, $\boxed{\textbf{SIN}}$ and $\boxed{\textbf{LN}}$ which are pressed prior to entering a numerical value.

- Multiplication and division

- Addition and subtraction

Exercises: Make sure that **Radian** mode is selected for trigonometric calculations.

1. Evaluate each expression taking care to place parentheses in the correct place so the desired result is obtained.

   (a) $\sin\frac{\pi}{3}$

   (b) $\dfrac{1}{2+\frac{3}{4+5}}$

   (c) $\dfrac{1}{3+\ln 2}$

   (d) $\cos^2\frac{\pi}{4} + \sin^2\frac{\pi}{4}$

   (e) $e^{5+5^{-1}}$

   (f) $27^{1/3}$

2. Compute $\frac{21.25-3.2}{16.3}+11.3$ and give the result in both decimal and fraction form.

3. Suppose there are 8 people who stand in line at the same bus stop every day. There are exactly 7 factorial (i.e. 7!) ways that these people can be ordered in line. If they agree to stand in line a different way every day, then how many years will it take for them to exhaust all possibilities?

4. Learn about the commands **lcm** and **gcd**. Find both the least common multiple and greatest common divisor of the integers 644 and 1376.

5. There was once a country with a terrible problem that was threatening its existence. Out of desperation, the ruler of this country offered to give a wonderful reward to any person who could solve this problem and help the country survive. Many people came forward with ideas, and finally there was a woman who solved the country's problem. When the ruler asked her to name her reward, she insisted that her request was a trivial one. She brought a checker board and told the ruler that she would return the next day and collect one grain of rice from the first square on the board. She would then return the next day to collect 2 grains of rice from the second square on the board. On each subsequent day (for 64 days) she would return to collect twice the number of grains of rice from the board as she collected on the previous day. Do you think she ever received her entire reward?

1.2   Storing and Recalling Variables

Often the result of a calculation will be used in later calculations. This is one reason why you might want to save a numerical value for later use. The TI-86 makes this easy to do using the $\boxed{\text{STO►}}$ and $\boxed{\text{2nd}}$ [RCL] buttons, which are the *store* and *recall* commands respectively. For identification purposes, a name is given to each value stored. Each name must begin with a letter and can be up to eight characters in length. Besides standard lower and upper case letters, Greek and other special characters are also available. The standard alphabet are printed in blue above keys on the right. The alphabetic characters are accessed by pressing $\boxed{\text{ALPHA}}$ for upper case and $\boxed{\text{2nd}}$ $\boxed{\text{ALPHA}}$ for lower case. The cursor will change from solid rectangle to a reverse video letter **A** and **a** respectively. After entering a letter from the keyboard the cursor reverts to the standard rectangle. If you want to type a longer string of letters, it is best to lock the alphabet keyboard by pressing $\boxed{\text{ALPHA}}$ a second time for either case. The alphabet keyboard is unlocked by again pressing $\boxed{\text{ALPHA}}$. When using the $\boxed{\text{STO►}}$ or $\boxed{\text{2nd}}$ [RCL] commands the locked alphabet keyboard is automatically set to make it easy for entering the name of the storage location. The following screen shot shows how an intermediate value is saved for a later calculation. The arrow symbol $\rightarrow$ is the result of pressing the $\boxed{\text{STO►}}$ key. Note that the variable name can be keyed in later calculations as part of an expression.

Furthermore, more than one variable can appear in an expression.

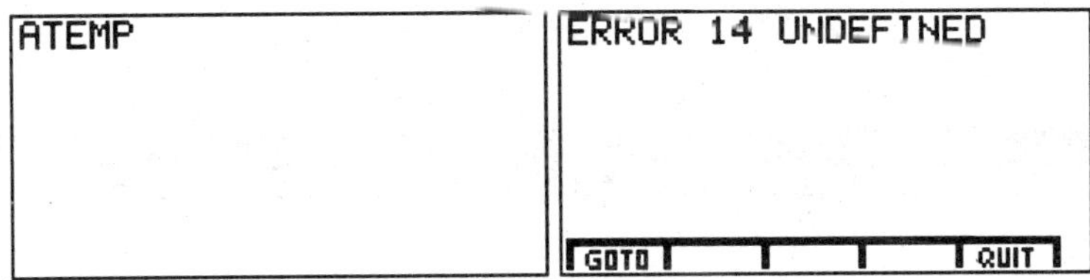

Once a result is stored in memory, it is kept in memory even after your calculator is turned off. The only way to lose these saved variables is to purposely delete these variables or have dead batteries.

There are a number of variables names which are used by the TI-86 as reserved names, and hence they can not be defined as new variables. These include the built-in command names, plus **Na, k, Cc, ec, Rc, Gc, g, Me, Mp, Mn, h, c, u,** π, **e**. A complete list is given in the **TI-86 Guidebook**.

Note that implied multiplication is a little tricky when two variables are multiplied. For example, if we try to multiply the variables **A** and **TEMP** using implied multiplication then we get an error message.

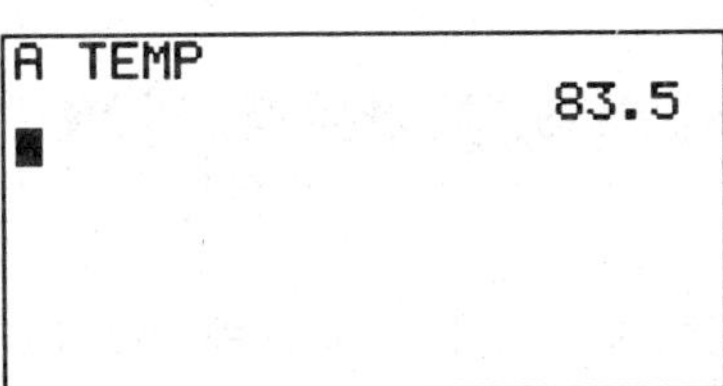

The reason is simple. The TI-86 is attempting to treat **ATEMP** as a variable name. Unfortunately, this variable name has not been assigned a value. If we place a space between **A** and **TEMP** then the TI-86 responds with the correct result.

The built-in *recall* mechanism can also be used to incorporate variables into expressions. When $\boxed{\textbf{2nd}}$ [RCL] is pressed, the screen shows the recall process on the eighth screen line, and the calculator is placed in alpha-lock mode. When the variable name is typed followed by the $\boxed{\textbf{ENTER}}$ key, the numerical value is substituted directly into the calculation.

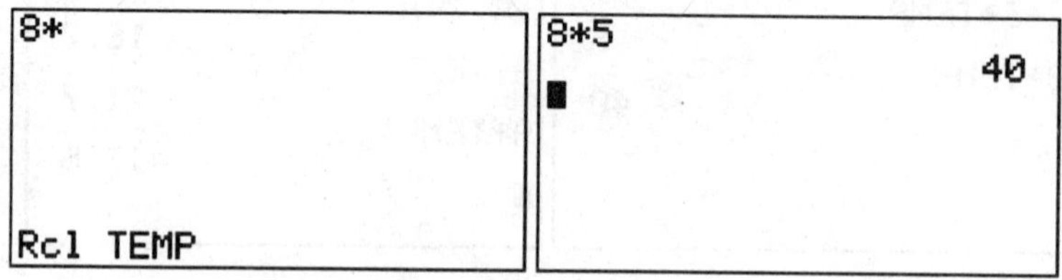

As mentioned above, Greek letters can be used in variable names. The Greek letters and other special characters are accessed by pressing 2nd [CHAR]. The first screen below shows the result of pressing 2nd [CHAR] followed by F2. This is the Greek letter menu which allows you to enter Greek letters as shown. Pressing MORE successively cycles through all the choices available. The second screen below shows the result of pressing MORE twice.

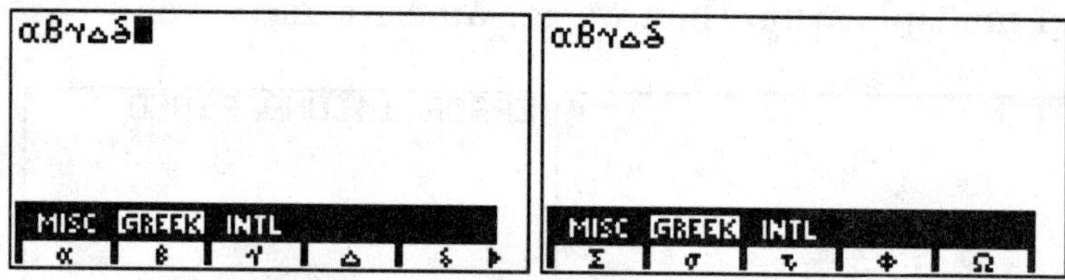

Using storage and recall with Greek letters is the same as for the standard letters.

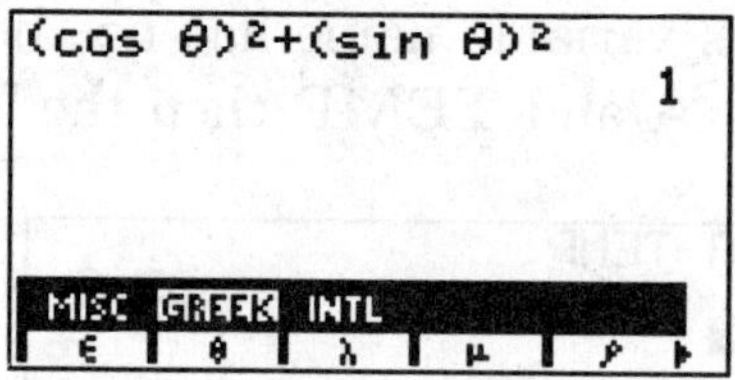

A complete listing of all assigned variables can be found by pressing 2nd [CATLG-VARS] (2nd [VARS] for the TI-85) to bring up the variables menu. The menu choices allow you to examine all variables, real variables only, complex variables only, program names, plus several other choices. Choose REAL to see the variables created in this section. Use the up/down arrow keys to select a variable and press ENTER to paste it to the home screen.

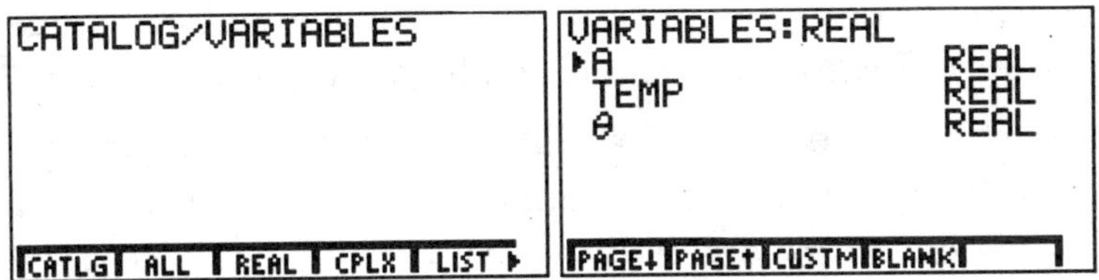

Memory Management. Finally, we should note that variables can be cleared from memory by using the TI-86's memory manager. For example, to clear one or more of the variables created in this section press $\boxed{\text{2nd}}$ [MEM] and select **DELETE** (press $\boxed{\text{F2}}$). Then select **REAL** (press $\boxed{\text{F2}}$).

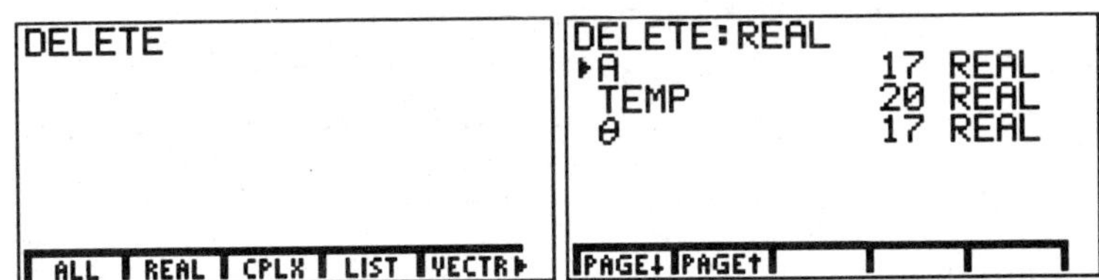

At this point, nothing has been deleted (and we can safely $\boxed{\text{EXIT}}$). However, if we want to delete the variable **TEMP**, then we can use the up/down arrow keys to select this entry and press $\boxed{\text{ENTER}}$.

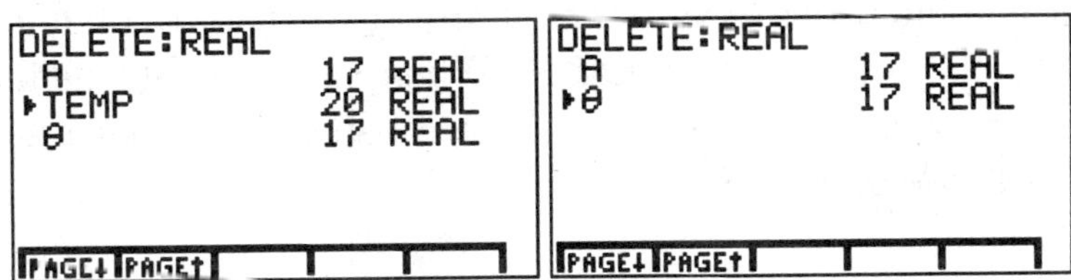

Press $\boxed{\text{EXIT}}$ to return to the home screen.

Exercises: Make sure that **Radian** mode is selected for trigonometric calculations.

1. Store the values 1.2, 1.35 and 2π in the variables **L**, **P** and **Q**. Now compute $\sin(L) - \cos(P) + c^Q$ and store the result in the variable **D**. What is the value of $D^3 - 4P^2 + 3Q$?

2. At the beginning of this section we noted that the variable names **Na**, **k**, **Cc**, **ec**, **Rc**, **Gc**, **g**, **Me**, **Mp**, **Mn**, **h**, **c**, **u**, π, and **e** are reserved by the calculator. Give the value of each of these constants and tell what it represents (look in chapter 4 of the **TI-86 Guidebook!!**).

3. Another way to define variables is by using equations. For example, we can store the value $\frac{7}{3}$ in the variable **A** by inputting **A=7/3** in the home screen and pressing $\boxed{\textbf{ENTER}}$. In addition, variables which are defined in this manner can be defined in terms of other variables. For example, we can store the variable **V**, denoting the volume of a right circular cone, by inputting $\textbf{V=}\frac{1}{3}\pi\textbf{R}^2\textbf{H}$, where **R** represents the radius of the base and **H** represents the height. The screens below show the computation of several values of **V**. Note that the colon allows us to enter multiple inputs on a single line and only see the result of the last calculation.

```
V=πR²H/3
                      Done
R=5:H=2.7:V
          70.6858347058
R=3:H=2.5:V
          23.5619449019
■
```

Now suppose

$$F = A^2 \sin(A + B) - 2\cos(A - B) + \frac{A}{A + B^2 + 1}$$

Use the process above to find the values of F corresponding to all possible combinations of values of A and B where $A \in \{1.1, 2.31, 4.1, 6\}$ and $B \in \{2.2, -1.5, 3.1\}$. **NOTE:** When variables are defined in this manner, they should be deleted before using the same names in the **SOLVER** (see section 1.4).

1.3   Introduction to Graphing

The graphing capabilities of the TI-86 are very amazing. The TI-86 can graph

- standard functions, $y = f(x)$

- functions defined parametrically, $[x(t), y(t)]$

- solutions to differential equations

- functions in polar coordinates, $r = g(\theta)$

In this section, we concentrate on graphs of standard functions. Parametric graphs and differential equation graphs are discussed in chapters 2 and 7 respectively. Related material can be found in Chapters 5 and 6 of the **TI-86 Guidebook**.

Function Plots: Let's start by considering a simple example. Suppose we want to graph the function

$$f(x) = \sin(x) + \frac{1}{2}\cos(3x)$$

on the interval $[-2\pi, 2\pi]$. There are four basic steps involved in this process.

1. Select the modes **Radian** and **Func**.

2. Set the dimensions for the graph window.

3. Define the function in the **y(x)=** window.

4. Select the **GRAPH** menu item.

Press $\boxed{\text{2nd}}$ [MODE] to select the **Radian** and **Func** modes. Then press $\boxed{\text{GRAPH}}$ and set the graph window dimensions by selecting the **WIND** menu item (press $\boxed{\text{F2}}$).

Then select **y(x)=** by pressing $\boxed{\text{F1}}$ and enter the function $f(x)$ given above. Finally, press $\boxed{\text{2nd}}$ [M5] to **GRAPH** the function.

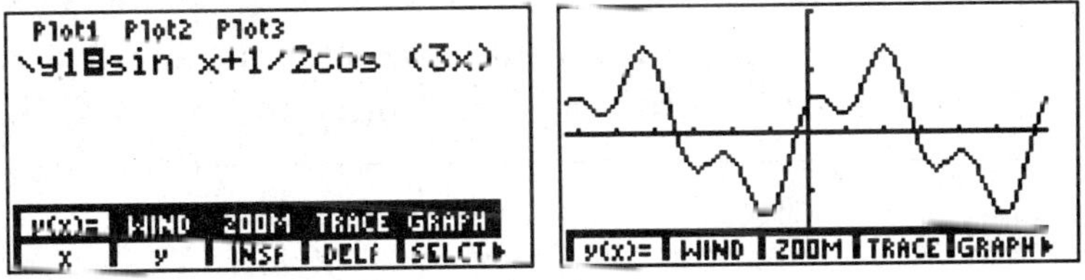

Using TRACE and ROOT: Several useful menu items are available from the second window above. We will investigate two of these here. Select

TRACE by pressing $\boxed{\text{F4}}$ and notice that cross hairs appear on the graph and coordinates appear at the bottom of the screen. The left/right arrow keys can be used to move the cursor along the curve. Notice that the coordinates change as the cursor moves along the graph.

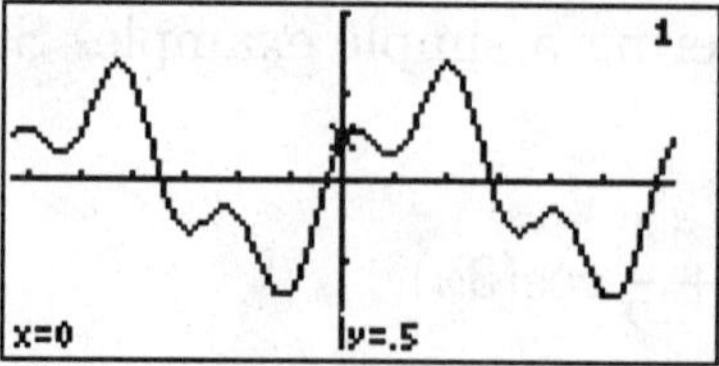

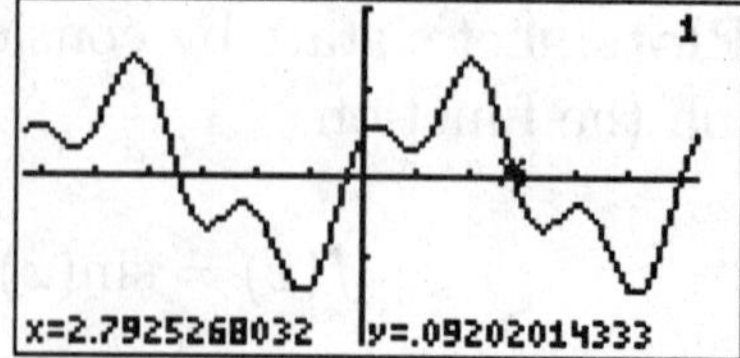

From the second picture above, there is a root near $x = 2.8$. We can find this root by pressing $\boxed{\text{EXIT}}$ (to reveal the menu items), pressing $\boxed{\text{MORE}}$, selecting **MATH** and selecting **ROOT**.

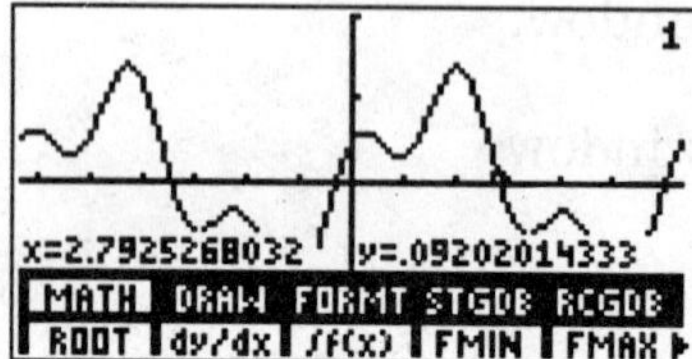

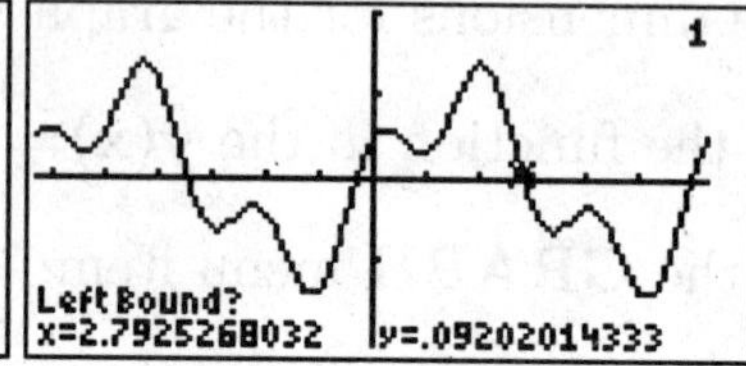

Then provide values for the left bound, the right bound and a guess. These can either be entered from the keyboard followed by $\boxed{\text{ENTER}}$ or by moving the cursor and pressing $\boxed{\text{ENTER}}$. The calculator will respond with the root and the corresponding value of y (which should be essentially zero).

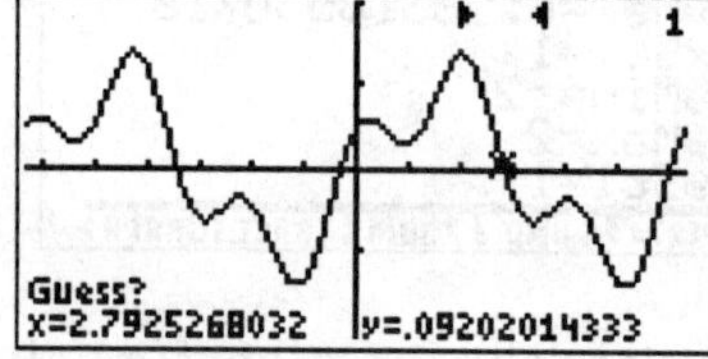

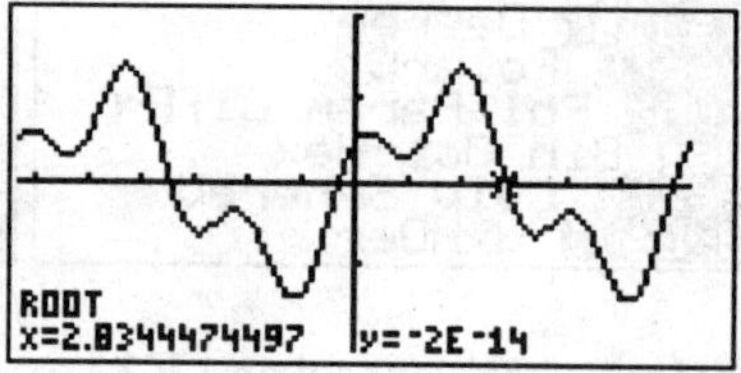

Finding the Intersection Point of Two Graphs: We can graph the function $g(x) = \cos(x)$ along with $f(x)$ by returning to the **y(x)=** screen (press $\boxed{\text{GRAPH}}$ and select **y(x)=**) and pressing the down arrow key (if necessary) to review the **y2=** prompt. Input $g(x)$ and select **GRAPH**.

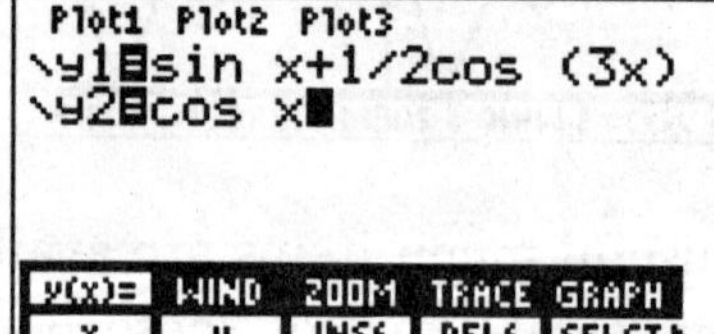

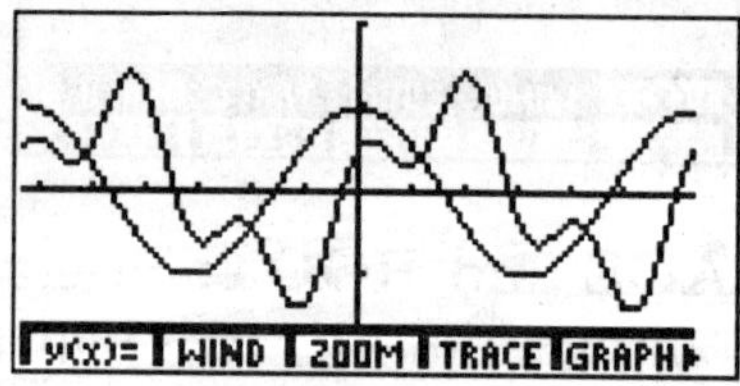

Notice that these graphs intersect at four points in this graph window. Let's find the first intersection point to the right of the y-axis. Select **TRACE** and move the cursor towards this point of intersection.

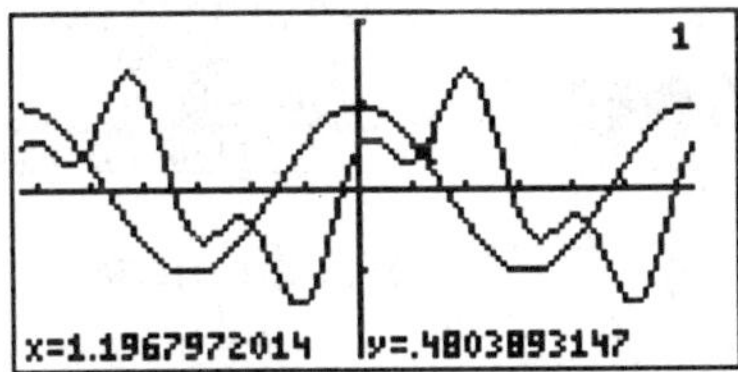

Then press $\boxed{\textbf{EXIT}}$ $\boxed{\textbf{MORE}}$ and select **MATH**. Finally, press $\boxed{\textbf{MORE}}$ and select **ISECT**.

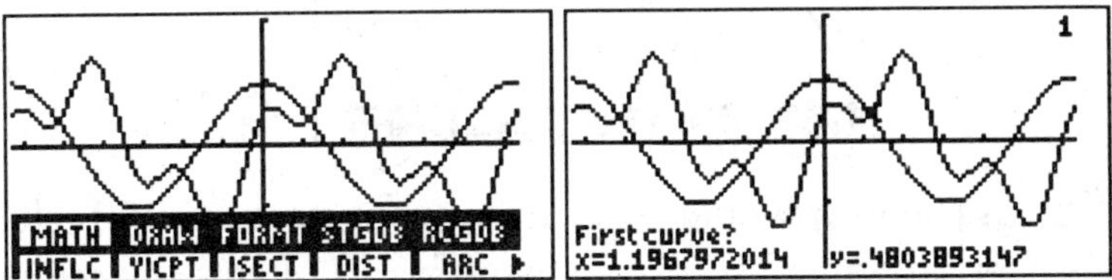

The TI-86 prompts for the first and second curve. Since there are only two curves present, simply press $\boxed{\textbf{ENTER}}$ following each prompt. Otherwise use the up/down arrow keys to select the curves (notice the number in the upper right had corner) and press $\boxed{\textbf{ENTER}}$. Then provide a guess by moving the cursor close to the intersection point and pressing $\boxed{\textbf{ENTER}}$.

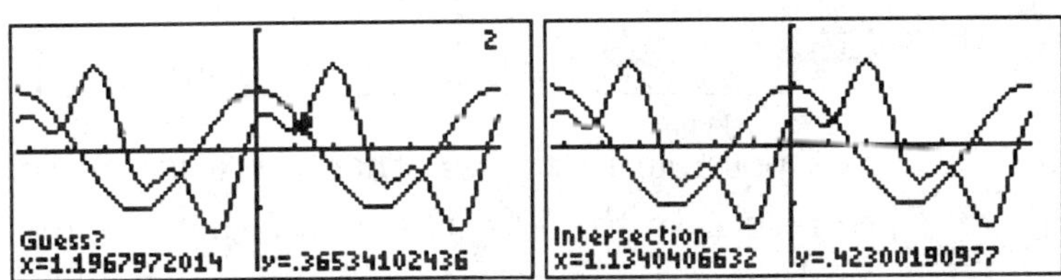

Using ZOOM: The **ZOOM** screen can be used to change the window dimensions so that a specific feature can be analyzed. Consider the function

$$f(x) = 10x^3 - 121x^2 + 221x - 110$$

Let's start by returning to the **y(x)=** screen and clearing both **y1** and **y2** (use the up/down arrow keys to select each item and press $\boxed{\textbf{CLEAR}}$, or select **DELf** (delete function). Then let's enter $f(x)$ as **y1** and set the window variables as shown below.

Notice that the formula for $f(x)$ does not fit in the window. The left/right arrow keys can be used to scroll the formula and check its accuracy. Finally, let's select **GRAPH**.

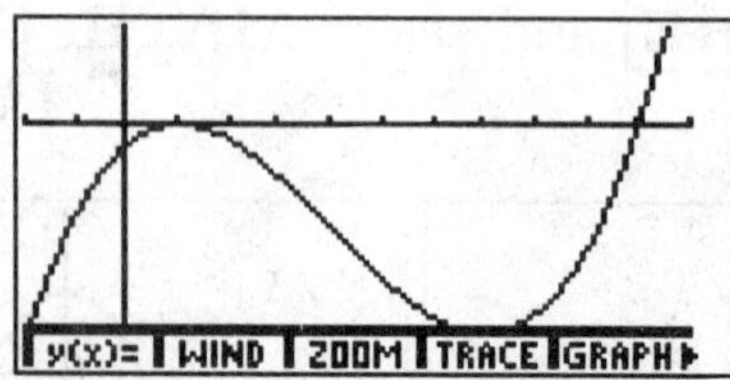

It is apparent from the graph that $f(x)$ has a root near $x = 10$. However, we cannot tell whether these is also a root near $x = 1$. Let's try zooming in on this part of the graph. Move the cursor to the portion of the graph that we need to see more clearly, select **ZOOM** and then **ZIN** (zoom in) followed by **ENTER**.

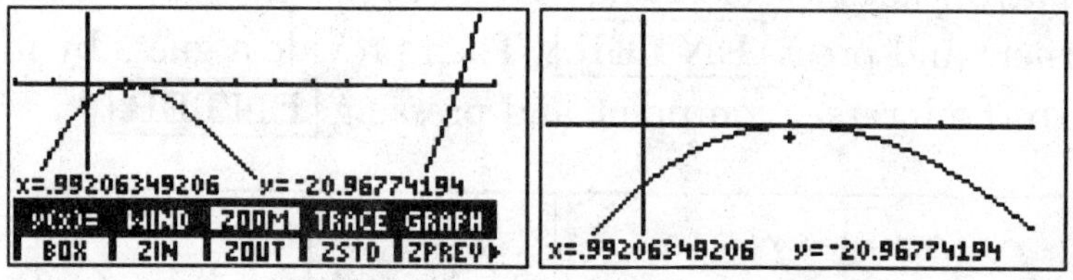

Notice that it is still not possible to see whether there is/are root(s) near $x = 1$. Press **EXIT** to return to the **ZOOM** menu. Let's refine the zoom process. Select **BOX** and move the cursor to the upper left hand corner of a rectangular region that you want to make the new graph window. Then press **ENTER**, move the cursor to the lower right hand corner, and press **ENTER**. The result of performing this process twice is shown below.

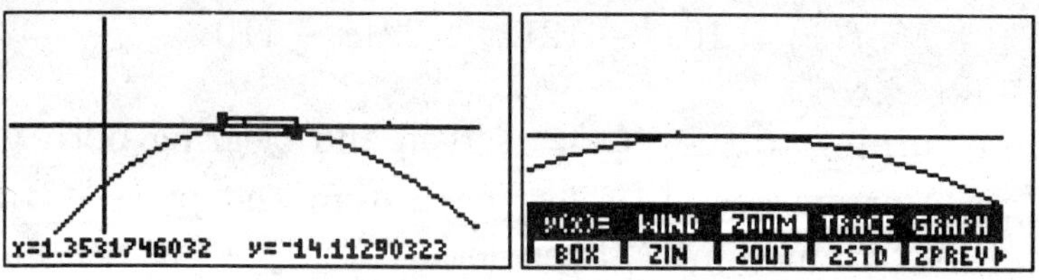

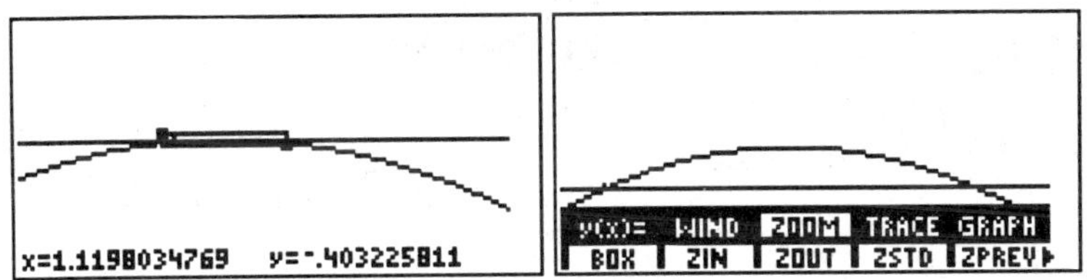

Now we can repeat the process of using **ROOT** from within the **MATH** menu (see above) to find the 2 roots.

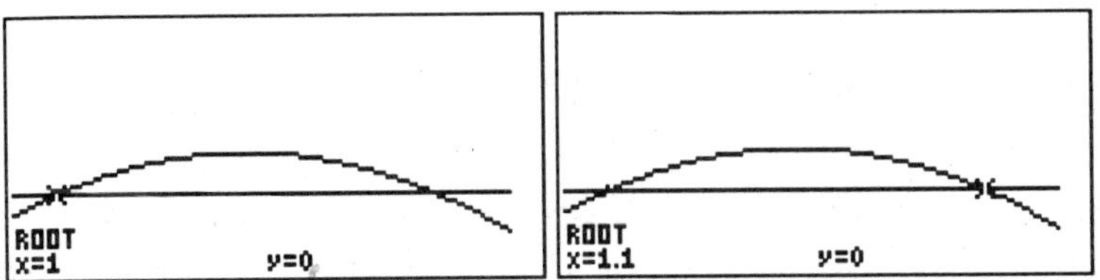

Exercises: Make sure that **Radian** mode is selected for trigonometric calculations.

1. How many solutions does the equation $\tan x = x$ have for $-10 \le x \le 10$. Use the **ISECT** command to find the solutions. How many solutions does this equation have for $-\infty < x < \infty$?

2. As an alternative approach to problem 1, consider the equation

$$\sin(x) - x\cos(x) = 0$$

   Does this equation have the same solutions as the previous equation? Use the **ROOT** command to find all of the solutions to this equation for $-20 \le x \le -10$.

3. Graph $f(x) = x\sin(x/(x^2 + 0.001))$ in a $[-1,1] \times [-1,1]$ window. Find all of the positive roots of $f(x)$ in this window. (Hint: $f(x)$ is an even function.)

4. Use a graph to find all zeros of the polynomial

$$f(x) = 5x^3 + 4x^2 - 34x - 45$$

   Is there a rational zero? If so, use it to help factor the polynomial.

5. Use graphs to verify the trigonometric identity

$$\sin(3x) = 4\sin(x)\cos^2(x) - \sin(x)$$

1.4   Using the Built-in Solvers

The TI-86 has three built-in solvers:

- **SOLVER** – For analyzing nonlinear equations in several variables

- **POLY** – For finding the roots of polynomials

- **SIMULT** – For solving linear systems of equations

We give a basic introduction to these solvers in this section. Additional information can be found in Chapter 15 of the **TI-86 Guidebook**.

Using the SOLVER: The **SOLVER** is a very nice feature for analyzing equations in several variables. For example, consider the law of cosines from trigonometry.

$$C^2 = A^2 + B^2 - 2AB\cos\theta$$

This formula is used to solve for an unknown side or angle in a triangle. Here A, B and C are the lengths of the sides of a triangle, and θ is the angle opposite the side of length C. Let's use the **SOLVER** to find the value of B given that $C = 4$, $A = 1$ and $\theta = \pi/12$. To start the solver, press $\boxed{\text{2nd}}$ [**SOLVER**] and enter the equation for the law of cosines (recall that the Greek letters are found by pressing $\boxed{\text{2nd}}$ [CHAR] and selecting **GREEK**) and press $\boxed{\textbf{ENTER}}$

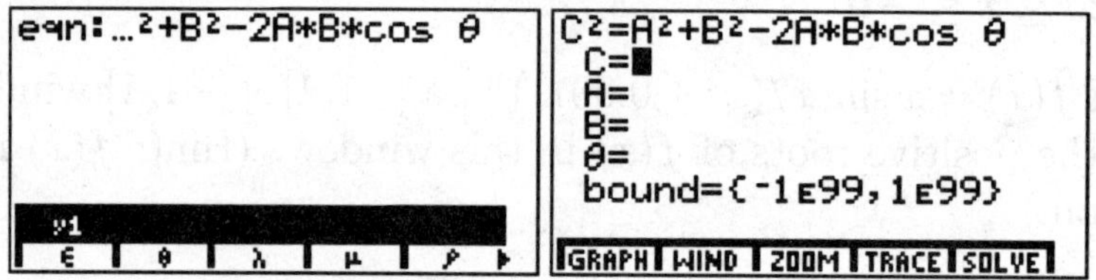

Notice that the TI-86 responds with a list of all variables in the equation along with a menu which contains several of the graphing commands. (**WARNING:** If any variable names which appear in the equation have previously been assigned values by using "=", then the **SOLVER** will not treat them as free variables.) Now, enter the values of the variables that you know, set the bounds (note that B must be positive) and then move the cursor to the B variable and select **SOLVE**.

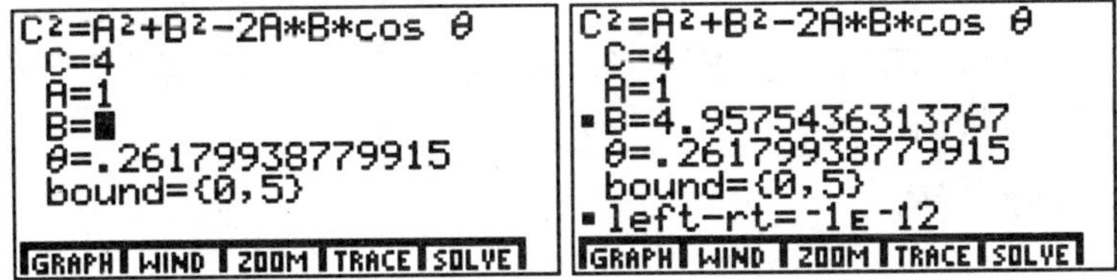

The value is calculated numerically, not algebraically, and the result is indicated with a bullet. The graph command plots a graph of the difference of the left and right sides of the equation as a function of the unknown variable using the current window variables. Plotting this graph can be helpful in deciding if there is more than one solution. Set the **xMin** and **xMax** variables to reflect the search interval and select **GRAPH**.

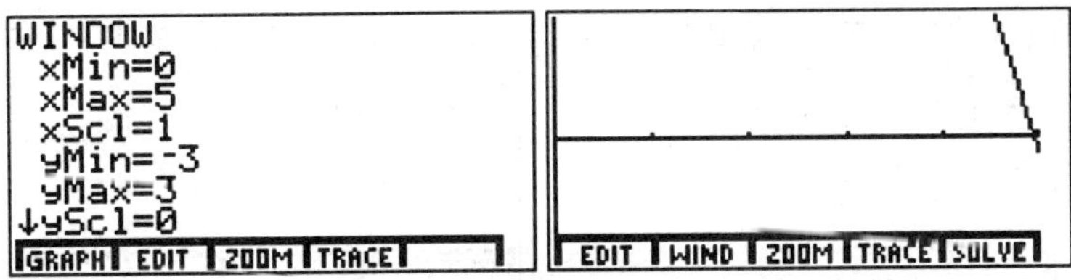

The **WIND, ZOOM** and **TRACE** menu items work in the same way as discussed in the previous section on graphing. This is especially useful when an equation has more than one solution for a particular variable.

Using POLY: The TI-86 has a built-in process for finding roots of polynomials. The process of finding the roots of the polynomial in the previous section is very simple using the polynomial root finder, **POLY**. Recall that we were trying to find the roots of the polynomial

$$f(x) = 10x^3 - 121x^2 + 221x - 110$$

Press $\boxed{\textbf{2nd}}$ [**POLY**], input the order (degree) of the polynomial and press $\boxed{\textbf{ENTER}}$.

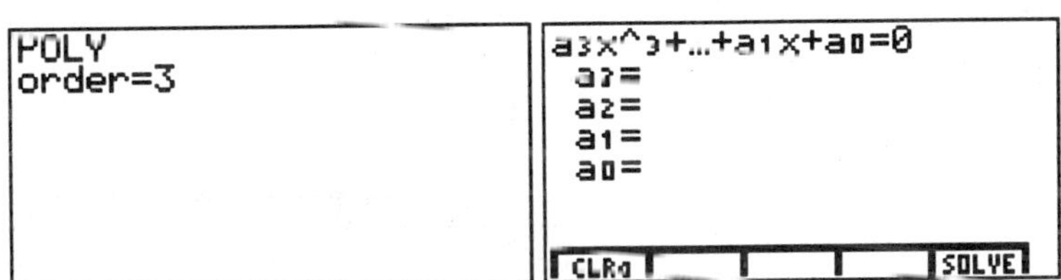

The TI-86 responds with a window including the equation and coefficient prompts. Enter the coefficient values and select **SOLVE**.

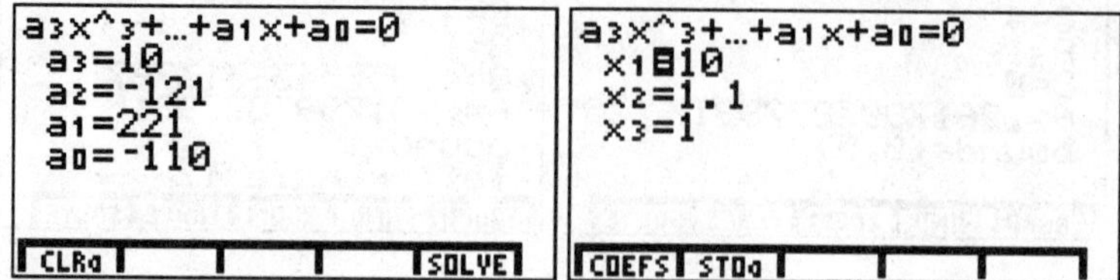

WOW! That's a lot easier than the process we used before!

Note that **POLY** will also give complex roots (if they exist). For example, if we use **POLY** to find the solutions of

$$x^3 - 4x^2 + 6x - 4 = 0$$

then we obtain

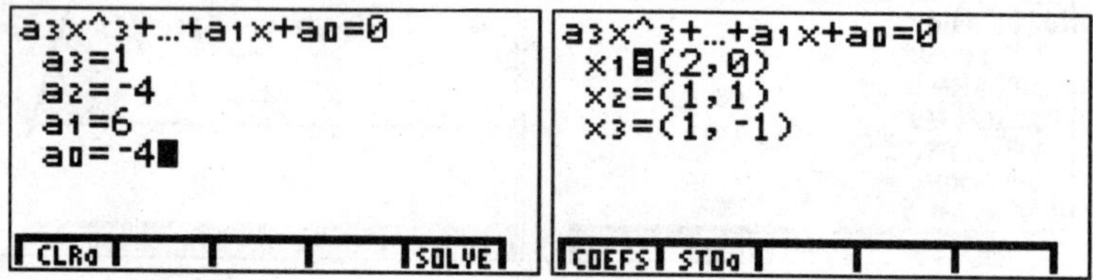

The TI-86 uses ordered pairs to represent complex numbers. Consequently, the response above indicates that the roots are 2, $1 + i$ and $1 - i$.

SIMULT: The final TI-86 solver is **SIMULT**, the solver for simultaneous systems of n linear equations involving n unknowns. For example, suppose we want the solution to the system

$$\left(\begin{array}{c} x + 2y - z = 5 \\ -2x + 3y + z = 2 \\ x - y - z = -1 \end{array} \right)$$

Press $\boxed{\textbf{2nd}}$ [**SIMULT**], input the number of equations (unknowns) and press $\boxed{\textbf{ENTER}}$. The TI-86 responds with a window requesting coefficient values.

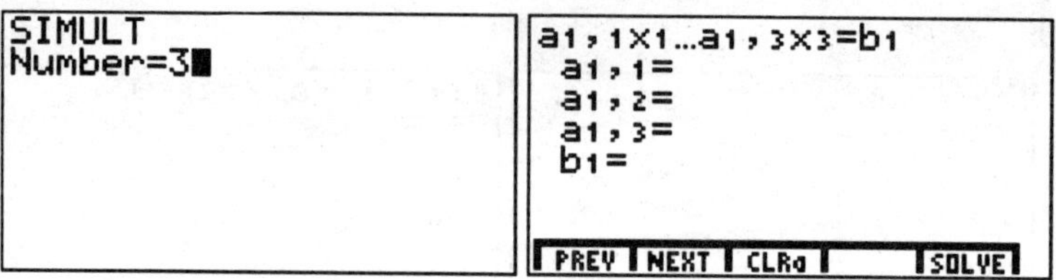

The value $a_{i,j}$ corresponds to the coefficient in the ith equation that multiplies by the jth variable, and the coefficient b_j corresponds to the value on the right hand side of the jth equation. For the system above, we have

$$a_{1,1} = 1 \quad a_{1,2} = 2 \quad a_{1,3} = -1 \quad b_1 = 5$$
$$a_{2,1} = -2 \quad a_{2,2} = 3 \quad a_{2,3} = 1 \quad b_2 = 2$$
$$a_{3,1} = 1 \quad a_{3,2} = -1 \quad a_{3,3} = -1 \quad b_3 = -1$$

Input these values and select **SOLVE**.

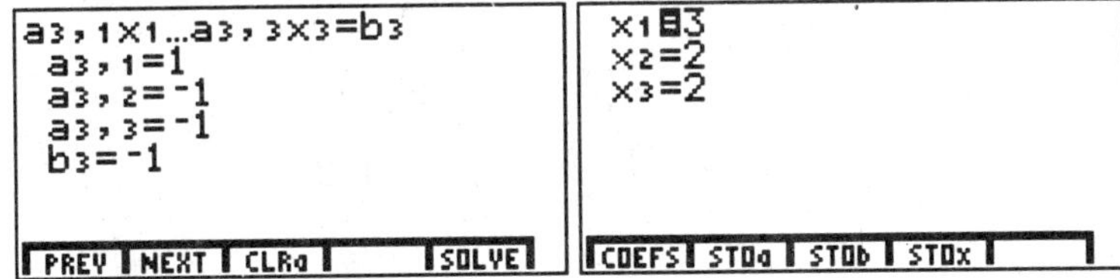

Consequently, the solution is given by $x = 3$, $y = 2$, $z = 2$.

NOTE: Linear systems of equations can have either 0, 1, or infinitely many solutions. If the system has either 0 or infinitely many solutions then the TI-86 will respond with an error message.

Exercises

1. If an amount of \$P is invested in a bank that pays an annual interest rate r compounded n times per year, then the value of the investment at time t is given by

$$F = P\left(1 + \frac{r}{n}\right)^{nt}$$

   (here r must be entered as a decimal; e.g. 6% is $r = 0.06$). Use the solver to set up this equation. Suppose \$100 is invested in a savings account that pays 5.5% compounded quarterly. In how many years will the account be worth \$200? What annual interest rate is required to triple \$100 in 15 years in an account that compounds interest quarterly?

2. If \$R is invested at the beginning of each month in an account that pays an annual interest rate r compounded monthly, then the amount of money in the account after n months is given by

$$S = R\left[\frac{12(1 + r/12)^{n+1} - 12 - r}{r}\right]$$

If \$100 is invested at the beginning of each month in an account that compounds monthly, then what interest rate is required for the account to have \$30,000 in 18 years? How about 40,000 in 18 years?

3. Set up the solver to accept arbitrary values of the coefficients and to solve the general quadratic equation

$$ax^2 + bx + c = 0.$$

Try to find the zeros for the following quadratic equations:

$$9x^2 - 66x + 121$$
$$77x^2 - 353x + 246$$

4. Repeat the previous exercise using the **POLY** solver.

5. Use the **POLY** solver to factor the polynomial

$$x^4 + \frac{17}{12}x^3 - \frac{13}{8}x^2 - \frac{2}{3}x + \frac{1}{2}$$

6. Use the **SIMULT** solver to solve

$$\left(\begin{array}{c} 2x - y + 3z = 7 \\ 3x + 5y + 7z = -2 \\ 5x - 2y + 2z = 5 \end{array} \right)$$

Give the solution in both decimal and fraction form.

7. Find the equation of the parabola that passes through the points $(-1, 1)$, $(1, 2)$, and $(2, 3)$.

8. Find the cubic polynomial $f(x)$ such that $f(1) = f(2) = f(3) = 1$ and $f(4) = 7$.

1.5   Lists and Tables

This section contains a basic introduction to lists and tables. Additional information can be found in Chapters 7 and 11 of the **TI-86 Guidebook**.

Suppose we are interested in values of the function

$$f(x) = 2\cos(x) - \sin(3x)$$

for $x = 1, 2, 3, 4, 5, 6, 7, 8$. One way to obtain these values is to enter $f(x)$ on the **y(x)=** screen as **y1** and then return to the home screen by pressing EXIT twice.

Now we can request values from the home screen. Simply press 2nd [CALC] and select **evalF** to paste this command to the home screen. Finally input the values shown below and press ENTER (recall that lower case letters are typed by pressing 2nd ALPHA).

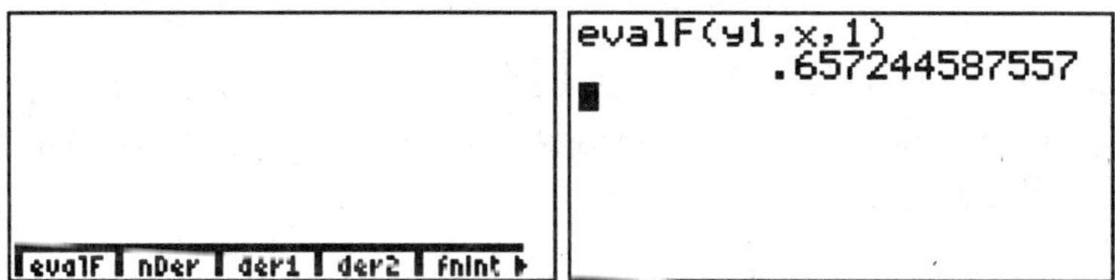

Now we can use 2nd [ENTRY] to find the other values by editing the third entry of **evalF** and pressing ENTER .

```
evalF(y1,x,1)
          .657244587557
evalF(y1,x,2)
          .005952821503
evalF(y1,x,3)
         -3.21634044893
```

Although this process works fine, it is much easier to find function values by using **Lists** and **Tables**.

Lists: Let's see how to use lists to compute function values. Consider the function $f(x)$ given above. We can use lists to evaluate $f(x)$ for $x = 1, 2, 3, 4, 5, 6, 7, 8$ by first creating a list of these values and giving it a name. Press 2nd [LIST] and select the "set brackets" from the menu items to create the list of x values and store it as **L**. Then use 2nd [ENTRY] twice to recall the **evalF(y1,x,3)** command from above, change the third entry to

L and press $\boxed{\textbf{ENTER}}$. We can use the right/left arrow keys to scroll the output.

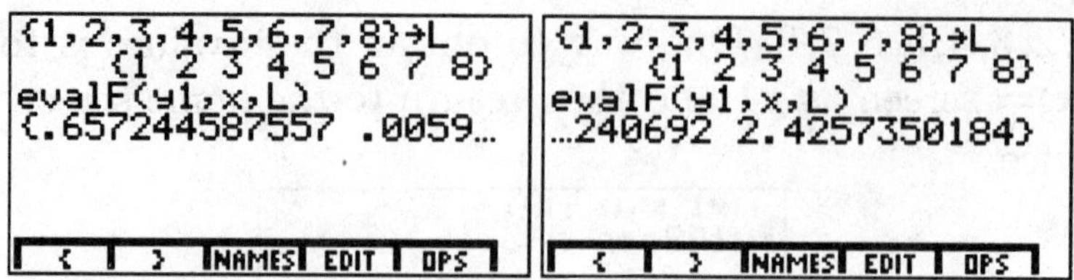

Now let's be a little fancy. We can create a list of the points corresponding to these values by entering the command line **L+(0,1)Ans** and store this as **PTS** as follows.

(This works because the TI-86 treats 1 as (1,0)...read about complex numbers and the TI-86.) Then we can request the entries in this list by entering **PTS(1)**, **PTS(2)**, etc.

Lists can also be used to graph families of functions. For example, if we want to graph the functions

$$g(x) = \sin(x)\cos(Kx)$$

for $K = 1, 2, 3, 4, 5$ then we can create the list of K values, place the formula for $g(x)$ in **y1**

set the window variables and select **GRAPH**.

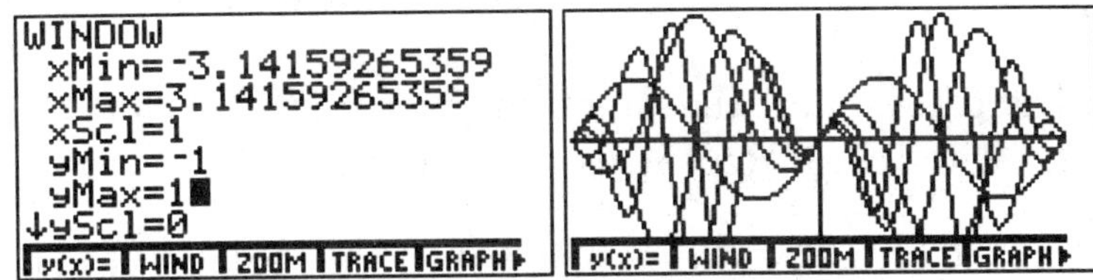

Tables: Perhaps the easiest way to analyze function values is to use a table. Consider the function $f(x) = 2\cos(x) - \sin(3x)$ from the beginning of this section. Input this function as **y1**. Then press $\boxed{\text{TABLE}}$ and select **TBLST** (table setup).

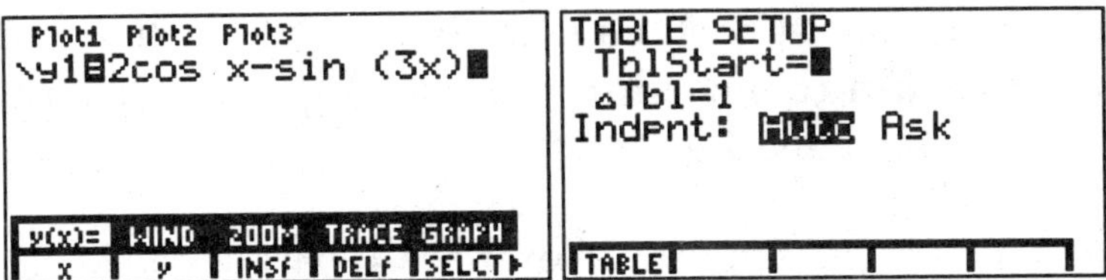

There are three important variables in this window. The **TblStart** variable determines the starting point for the free variable, the **ΔTbl** variable deter mines the difference between free variable values, and the setting for **Indpnt** determines whether the independent variable values are pre-set or if the table asks for values (try this). Set the values shown in the first window below and select the **TABLE** menu item.

The values in the **y1** column are the function values corresponding to the values of **x** in the first column. Pressing the up/down arrow keys scrolls through values indefinitely.

Exercises: Make sure that **Radian** mode is selected for trigonometric calculations.

1. Use a list to plot $y = x^n$ for $n = 1, 2, 3, 4, 5, 8, 11$ in a $[-1, 1] \times [-1, 1]$ window.

2. Use a list to plot $y = \sin(k\pi x)$ for $k = 1, 2, 3, 4$ in a $[0, 2] \times [-1, 1]$ window.

3. Use a list to evaluate the function $f(x) = x^2 - \sin(x)$ at the values $x = -2, -1, 0, 1, 2, 3$. Repeat using a table instead.

4. Store the list $\{1, 2, 3, 4, 5, 6, 7, 8, 9, 10\}$ in the variable **L**. Then use **L** to do the following:

   (a) Compute a list containing k^2 for $k = 1, 2, ..., 10$ (try **L^2**). Then use the **prod** command (found by pressing $\boxed{\textbf{2nd}}$ [**LIST**], selecting **OPS** and pressing $\boxed{\textbf{MORE}}$) to compute $2^2 3^2 \cdots 10^2$.

   (b) Compute the list containing 2^k for $k = 1, 2, ..., 10$ (try **2^L**). Then use the **sum** command to compute $\sum_{k=1}^{10} 2^k$. Give the answer in both decimal and fraction form.

5. The **seq( )** function (found as in exercise 4, part a) can be used to create lists in which the kth entry is given by some function of k. For example, entering **seq(2k-5,k,1,6)** produces the list $\{-3, -1, 1, 3, 5, 7\}$. Use **seq( )** to rework problem 5.

6. Use a table with the **Indpnt** mode set to **Ask** in **TBLSET** to approximate the maximum value of the function $g(x) = 1 + 3x - x^4$. Give an explanation (using complete sentences) of the strategy that you employ.

7. Repeat exercise 6 using a graph and the **FMAX** command from within the **MATH** menu accessed from the graph window.

8. Suppose $f(x) = \sin(x) + \frac{1}{2}\cos(3x)$. Use a table to evaluate this function at $x = \frac{k}{4}$ for $k = 1, 2, 3, 4, 5, 6, 7, 8, 9, 10$.

Chapter 2

Graphing Functions and Equations

This chapter introduces a number of techniques for defining and plotting functions and parametric equations. Additional information can be found in Chapters 5, 6 and 9 of the **TI-86 Guidebook**, and related material can be found in Chapter 1 of Stewart's **Calculus**.

2.1 Functions

A function can be defined in two ways on the TI-86. In section 1.3 we defined functions by using the **y(x)=** menu from the $\boxed{\textbf{GRAPH}}$ screen. It is also possible to define functions from the home screen.

Defining Functions From the Home Screen: For example, we can define the function

$$f(x) = 10 - x^2$$

as shown in the first screen below by using the $\boxed{\textbf{2nd}}$ $\boxed{\textbf{ALPHA}}$ [Y] keys for **y** and the $\boxed{\textbf{x-VAR}}$ key for x. We can also evaluate this function at $x = 3.1$ by selecting **evalF** from the $\boxed{\textbf{2nd}}$ [CALC] menu.

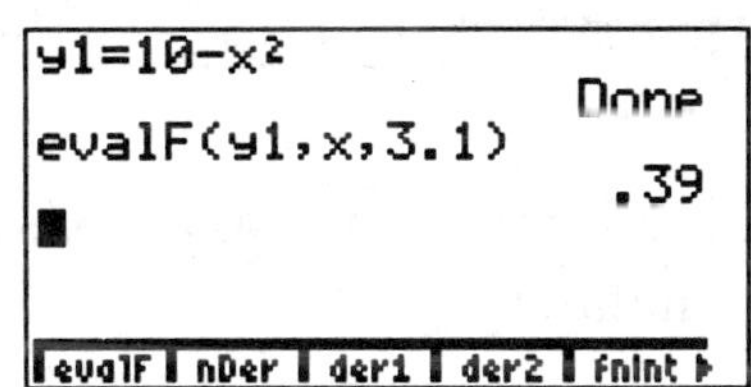

Now press $\boxed{\textbf{GRAPH}}$ and select **y(x)=**. Notice that $f(x)$ has been input as **y1**.

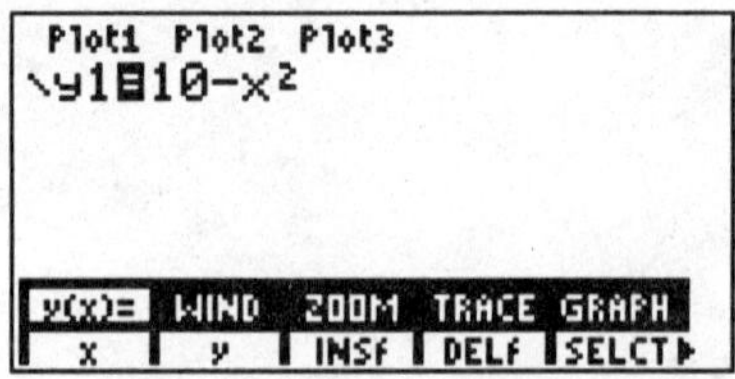

Let's graph this function in a $[-4, 4] \times [-10, 11]$ graph window.

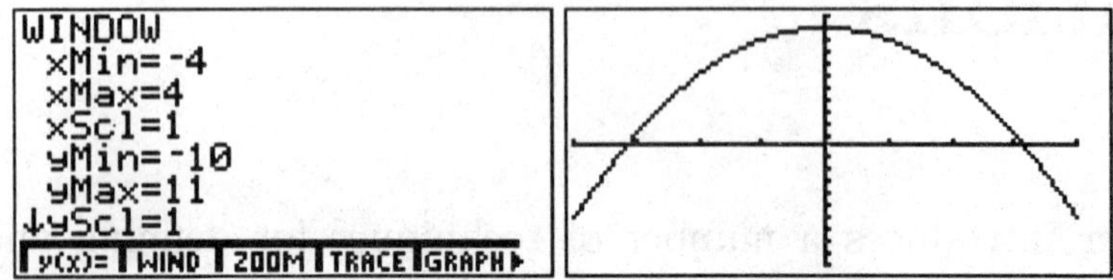

(Note: Press $\boxed{\textbf{CLEAR}}$ to remove the menus from the graph window. The menus can be restored by pressing either $\boxed{\textbf{GRAPH}}$ or $\boxed{\textbf{EXIT}}$.) Once the function is graphed, evaluation of the function is possible using either the **TRACE** or **EVAL** commands. We discussed the **TRACE** command in section 1.3, and we saw that **TRACE** only allows the function to be evaluated at pixel values. The **EVAL** command (short for evaluation) makes it possible to evaluate $f(x)$ at specific values of x. The **EVAL** command is accessed from the graph window by pressing $\boxed{\textbf{MORE}}$ twice and then selecting **EVAL**. A prompt appears requesting a value for x. Enter the value and press $\boxed{\textbf{ENTER}}$ to obtain the function value.

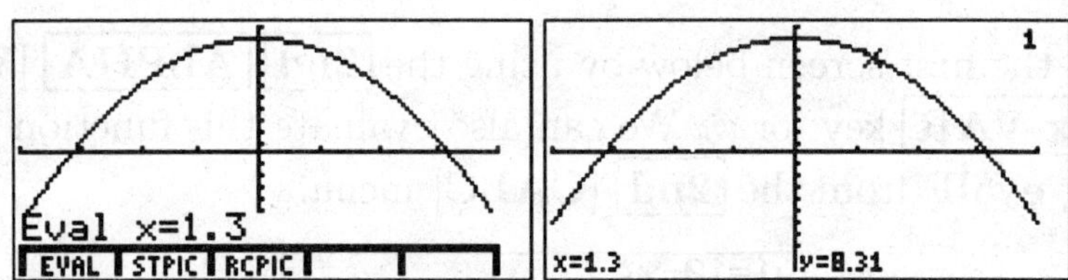

If you want to do several more function evaluations press $\boxed{\textbf{EXIT}}$ and then again select the **EVAL** command.

WARNING: If a we input a value for x which is not in the current graph window then we get an error message.

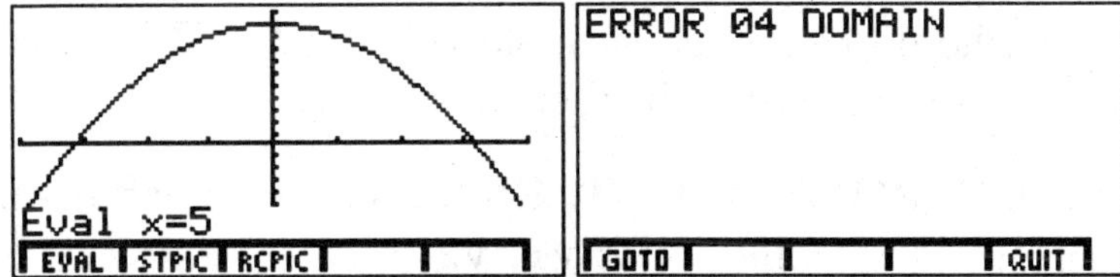

One way to avoid the problem mentioned above it to define new window dimensions. We can do this manually, or by using the **TRACE** command. The **TRACE** command can be used to interactively move the graph window. To see this, select **TRACE** and scroll the cursor towards the edge of the graph window.

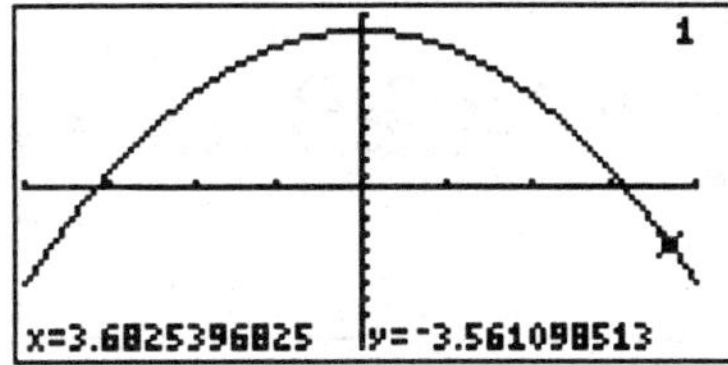

Now continue to scroll to the right. Notice how the graph window re-sizes itself in the x direction (but not the y direction). However, scrolling will continue even after the cursor has left the screen.

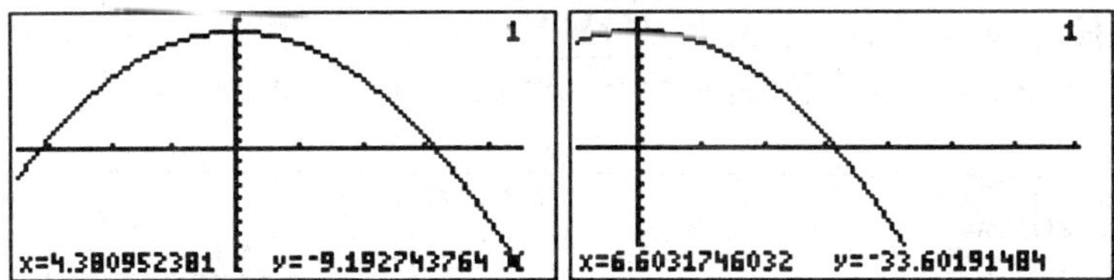

If you want the new graph window to be centered (both in x and y) at the cursor, press ENTER. Select **TRACE** to resume tracing. If we look at the current window dimensions (select **WIND**) then we see the settings have changed.

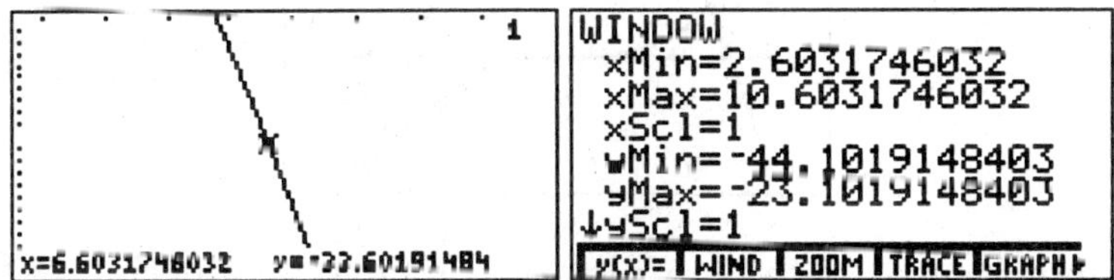

Functions can also be defined using other independent and dependent variable names. However, there are some limitations.

- Functions can only be graphed if their independent variable is x.

- Functions whose dependent variable is not **y1, y2,** etc. can not be plotted if lists are included in their definition.

- The **TRACE, ROOT, ISECT, EVAL,** etc. commands will not work if the dependent variable is not **y1, y2,** etc.

Even with these limitations, there are times when it is useful to define a function using other variable names. If we define $f(x) = \sin(x) - x$ as **F** (shown below), then we can still obtain function values from the home screen by using **evalF**.

We can also obtain a graph by using the **DrawF** command. The **DrawF** command can be accessed by either pressing $\boxed{\textbf{2nd}}$ [**CATLG-VARS**], selecting **CATLG** and scrolling to **DrawF**, or by pressing $\boxed{\textbf{GRAPH}}\ \boxed{\textbf{MORE}}$, selecting **DRAW** and pressing $\boxed{\textbf{MORE}}$ (or by simply typing it on the home screen). Note that **DrawF** uses the current graph window dimensions.

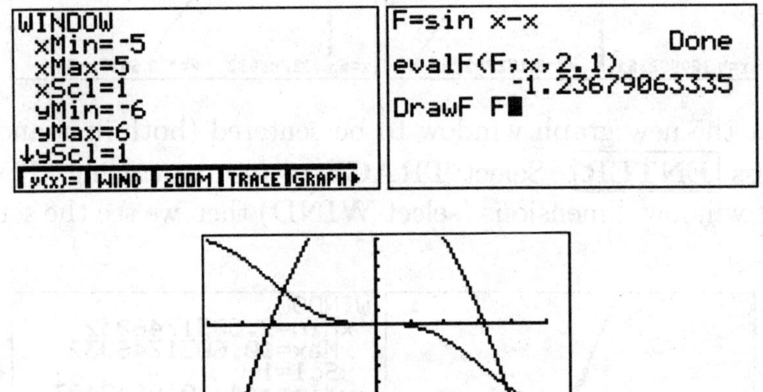

The plot window also contains a plot of the current function defined in the **y(x)=** screen.

Selecting and Shading Graphs: We conclude this section with some comments related to the **y(x)=** menu items. Press $\boxed{\textbf{GRAPH}}$ and use the

$y(x)=$ and **WIND** menu items to input the values shown below.

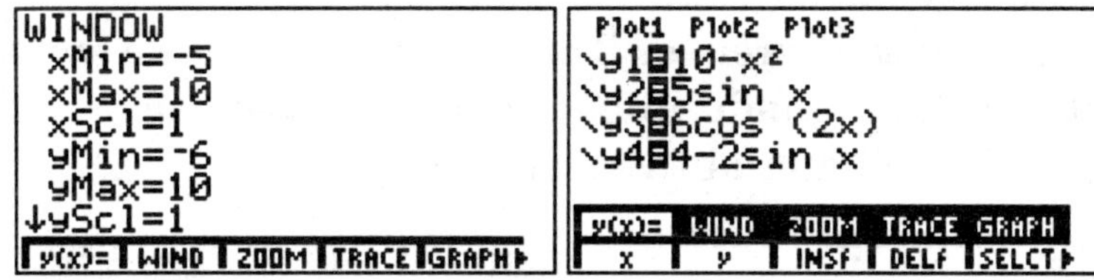

If we select **GRAPH** then we see the mess below.

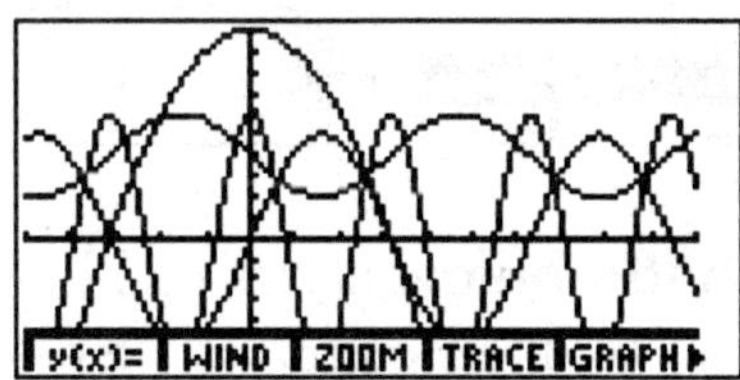

It is possible to restrict the graph window to a subset of the functions **y1, y2, y3, y4** without deleting these functions from the $y(x)=$ screen. Simply return to the $y(x)=$ screen, move the cursor to the **y1** equation and press **SELCT**.

Notice that the "=" sign is no longer highlighted. Pressing **SELCT** repeatedly toggles the highlighting of the "=" sign. The TI-86 will only graph the functions whose "=" signs are highlighted. Consequently, if we only want to see **y2** and **y4** then we use **SELCT** to obtain the first screen below and then graph the functions.

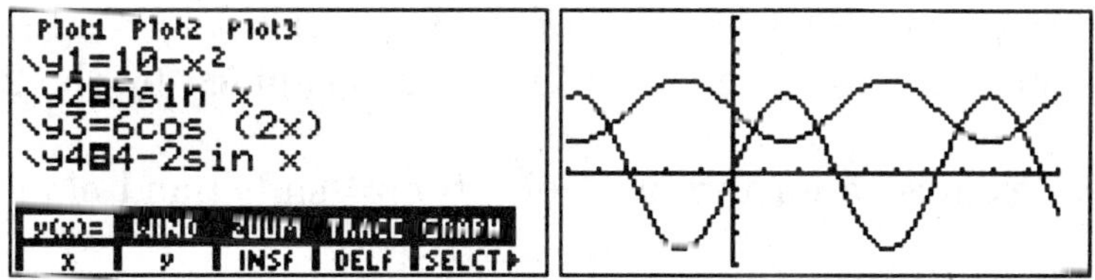

(The menu items have been removed from the plot by pressing $\boxed{\textbf{CLEAR}}$.)

It is also possible to partially shade graphs from the $y(x)=$ screen. Return to the $y(x)=$ screen and use **SELCT** to select **y1** (and "turn off" **y2**,

y3, y4). Then move the cursor next to the **y1** equation, press $\boxed{\textbf{MORE}}$ and toggle the style menu item. Notice how the style indicator to the right of **y1** changes as this button is pressed. There are seven possible styles, including **dot, line, thick line, shade above, shade below, path** and **animate**. We demonstrate the **shade above** style below.

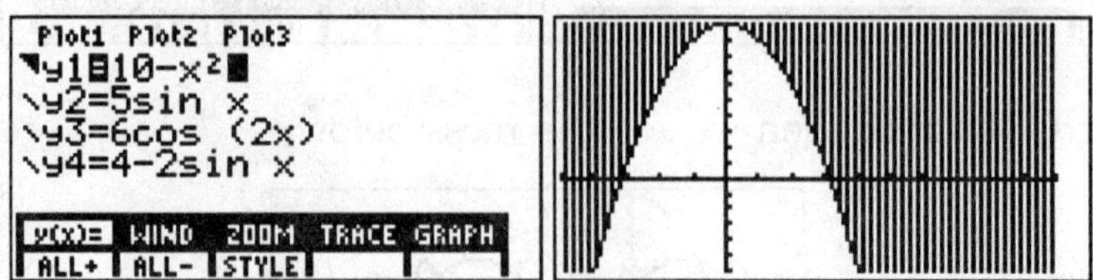

The **shade above** and **shade below** styles can be very useful for visualizing inequalities and area. The built-in command **Shade** can be used to create additional shading effects. See Chapter 6 in the **TI-86 Guidebook** for additional information.

Exercises: Learn about the **FMIN** and **FMAX** commands before attempting exercises 1-4. Learn about the **Shade** command before attempting exercises 5-8.

1. Find all of the zeros, local minimum values and local maximum values for the function $f(x) = x^3 - 3x^2 + x + 2$ on $[-1, 3]$.

2. Repeat exercise 1 for $f(x) = \sin(x^2) - \sin^2(x)$ on $[0, 3]$.

3. Repeat exercise 1 for $f(x) = e^{-x}\cos(\pi x)$ on $[-0.5, 2.5]$.

4. Repeat exercise 1 for $f(x) = x^6 - 2x^3$ on $[-1, 1.5]$.

5. Find the points of intersection of the functions $f(x) = x^3 - 3x^2 + x + 2$ and $g(x) = x + 2$. Then shade the region bounded by the 2 curves.

6. Repeat exercise 5 for the functions $f(x) = \sin(x)$ and $g(x) = x^2$.

7. Repeat exercise 5 for the functions $f(x) = (x+1)^2(x-1)^2$ and $g(x) = 2 - (x+1)^2(x-1)^2$.

8. Repeat exercise 5 for the functions $f(x) = e^{-x}$ and $g(x) = x(2-x)$.

2.2 New Functions from Old Functions

The TI-86 has a wealth of built-in mathematical functions. Aside from the functions which appear on the key board, some of the most important ones include:

- absolute value – $\boxed{\textbf{2nd}}$ [**MATH**] **NUM abs**

- n^{th} root – $\boxed{\textbf{2nd}}$ [**MATH**] **MISC** $\boxed{\textbf{MORE}}$ $^x\!\sqrt{}$

- hyperbolic and inverse hyperbolic functions – $\boxed{\textbf{2nd}}$ [**MATH**] **HYP** (make a selection)

Function Composition: TI-86 functions can be combined to create new functions through composition and algebraic combination. For example, consider the calculator screen shown below. The **evalF** command has been used in **y3** to create the composition of **y2** with **y1**. Similarly, **y4** is the composition of **y1** with **y2**. Algebraic combinations can also be formed, as **y5** shows.

```
Plot1  Plot2  Plot3
\y1=10-x
\y2=x²
\y3█evalF(y2,x,y1)
\y4█evalF(y1,x,y2)
\y5=ln (y2+1)-2 y3█
 y(x)=  WIND  ZOOM  TRACE  GRAPH
   x      y    INSf  DELf  SELCT▶
```

The plots of **y3** and **y4** are shown below.

```
WINDOW
 xMin=0
 xMax=20
 xScl=1
 yMin=-100
 yMax=100
↓yScl=10
 y(x)=  WIND  ZOOM  TRACE  GRAPH▶
```

As expected, the compositions are not the same.

The luxury of combining functions in this way makes it easy to display complicated functions. For example, if

$$F(x) = \sin\left(\frac{x^2 + 7x - 9}{x^3 + 1}\right)$$

then we can create this function piece-by-piece as shown below.

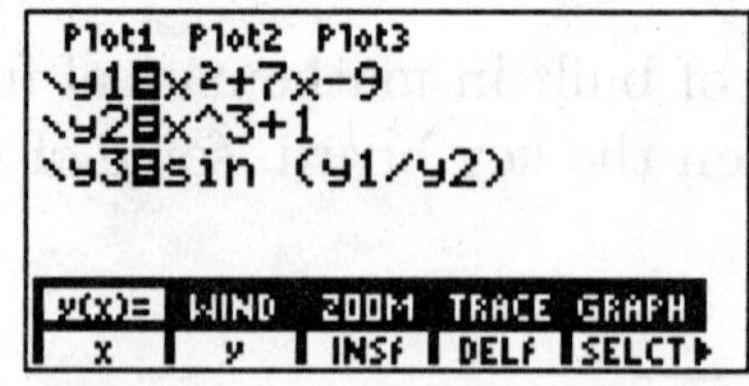

Piecewise Defined Functions: A piecewise defined function is one which is defined by different equations in different parts of its domain. For example,

$$f(x) = \begin{cases} 2 & \text{if } x < 3 \\ x - 1 & \text{if } x \geq 3 \end{cases}$$

gives two rules for different parts of the domain. One way to graph this function is to graph each rule as a separate function for the appropriate window size and then sketch on a piece of paper the graph for the complete function. This is not very satisfactory, but with a little bit of ingenuity, the correct graph of this function and ones like it can be done with the TI-86 calculator. A better way to graph piecewise defined functions is to use the **TEST** operators.

To access the **TEST** operators press $\boxed{\text{2nd}}$ [TEST]. The six choices are $==, <, >, \leq, \geq$ and $\neq$. The double equal signs are used to distinguish from the standard equals sign that is used in function definitions and equations defined in the **SOLVER**. These six tests allow you to compare two numbers. If the statement is true, the result is 1 and if the statement is false, the result is zero. The screen shown illustrates how the test is used.

The variable A is set to 5. Then it is compared with the number 4. The first inequality is true so the output is 1, and the second inequality is false so the output is 0.

Now let's return to the function $f(x)$ defined above. We can use the test operators $<$ and $\geq$ to define $f(x)$ as

$$f(x) = 2(x < 3) + (x \geq 3)(x - 1).$$

Note that when the first test is true, the second test is false. Consequently, $f(x) = 2$ when $x < 3$. Similarly, when the second test is true, the first is false. So $f(x) = x - 2$ when $x \geq 3$. The function is input as **y1** below, and the graph is shown on a $[-1, 6] \times [-1, 6]$ graph window.

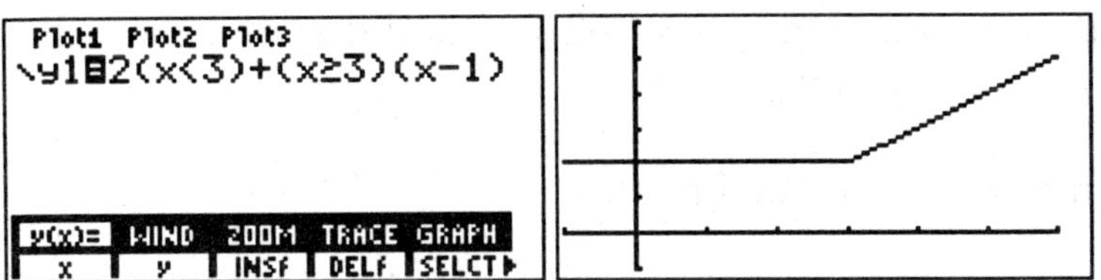

All the **ZOOM** and **TRACE** features are still valid for use with piecewise defined functions.

Now let's consider a piecewise function defined over three intervals. More specifically, suppose

$$g(x) = \begin{cases} x + 2 & \text{if } x \leq -1 \\ x^2 & \text{if } -1 < x < 1 \\ 1 & \text{if } x \geq 1 \end{cases}.$$

Using the test operations we can rewrite $g(x)$ as

$$g(x) = (x \leq -1)(x + 2) + (-1 < x)(x < 1)x^2 + (x \geq 1).$$

Note that for each value of x, only one of $(x \leq -1)$, $(-1 < x)(x < 1)$ and $(x \geq 1)$ is 1 (while the rest will be 0). The graph of $g(x)$ is shown at the top of the next page in a $[-5, 6] \times [-3, 3]$ graph window.

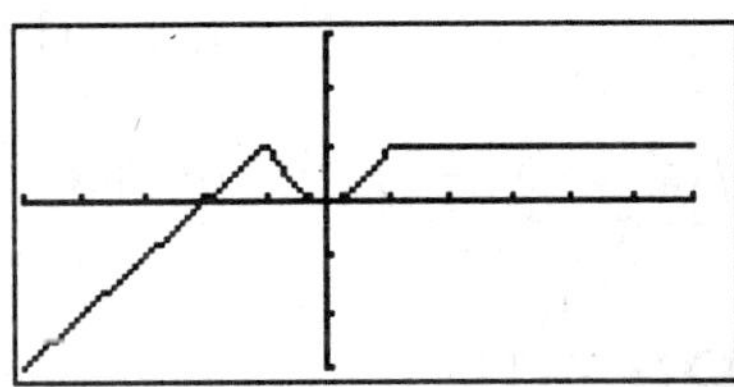

Exercises

1. Plot the function

$$f(x) = \begin{cases} x^2 & \text{if } x < 0 \\ 3x^2 & \text{if } x \geq 0 \end{cases}$$

Then, evaluate this function at $x = -2, 0, 2$ and zoom into the origin to examine how the pieces fit together.

2. Plot the function

$$f(x) = \begin{cases} \cos(2\pi x) & \text{if } |2| < 2 \\ -3\,|x| + 7 & \text{if } |x| \geq 2 \end{cases}$$

Evaluate this function at $x = -2, 0, 2$ and describe the behavior of the function near $x = -2, 2$.

3. Explain how to plot the function

$$g(x) = \begin{cases} \sin(x)/x & \text{for } x < 0 \\ x + 1 & \text{for } 0 \leq x < 2 \\ 3 - x^2 & \text{for } x \geq 2 \end{cases}$$

on the TI-86. Use the zoom feature to describe the behavior of $g(x)$ near $x = 0$.

4. Define the functions $f(x) = x^2 + 5x + 6$ and $g(x) = x^3 - 1$. Plot $f(x)/g(x)$, $\sin(g(x))$, $f(g(x))$ and $g(f(x))$.

5. Enter the function **x^2/(x$\geq$ −1)/(x$\leq$1)** as **y1** and give the graph in a $[-2, 2] \times [-1, 2]$ graph window with **AxesOff** (select **FORMT** from within the $\boxed{\textbf{GRAPH}}$ menu). Explain the result

6. Learn about the **int** command (found by pressing $\boxed{\textbf{2nd}}$ [**MATH**] **NUM**). Graph the function $f(x) = x - int(x)$ on the interval $[-2, 2]$.

7. Graph the function $f(x) = (x - int(x))^2$ on the interval $[-2, 2]$.

8. Graph the function $f(x) = x(-1)^{int(x)}$ on the interval $[-4, 4]$.

2.3   Graphing Pitfalls

The primary pitfall in using a graphing calculator occurs when poor dimensions are chosen for the graph window. A poor choice of dimensions can result in only seeing part of the graph, none of the graph, or all of the graph with some of the features obscured. For example consider the simple example of the function

$$f(x) = x^2 + 2.$$

If we choose the window dimensions to be $[-4, 4] \times [-4, 3]$, we only see part of the graph, and the majority of the display screen is not used.

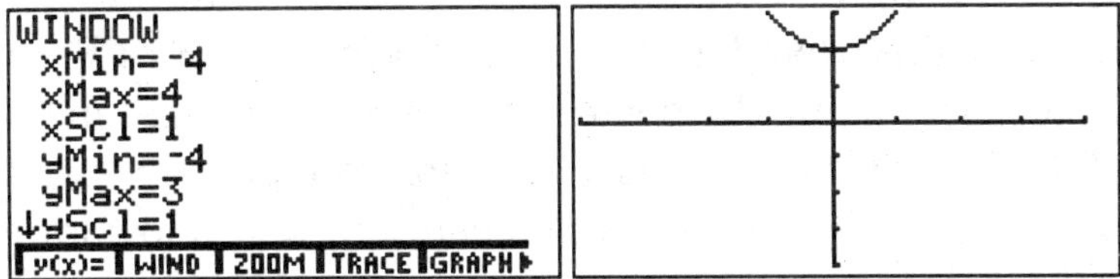

Even worse, if we choose $[-4, 4] \times [-2, 2]$, we see a blank screen. A choice which looks good at first glance is $[-15, 15] \times [-5, 200]$. In this case we get

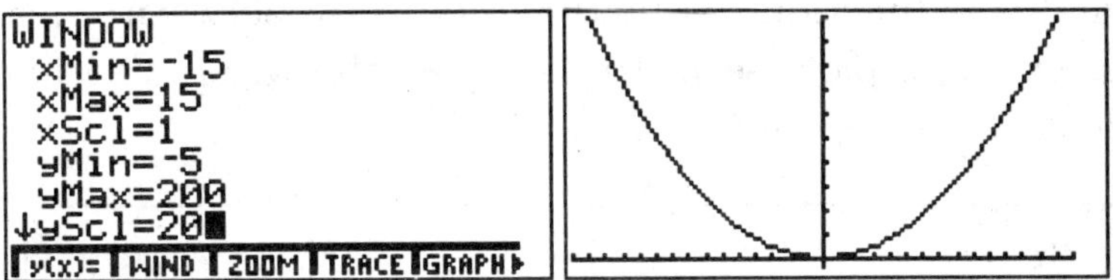

In this case we get a fairly complete graph, but it obscures the fact that the y-intercept is at $y = 2$, and not the origin.

The best approach to choosing a viewing window is to think about the domain and range for the function. The domain can be calculated by hand for most functions encountered in a calculus class. Once you know the domain, choose any values for **xMin** and **xMax** that are contained in the domain. The range is not always easy to find with pencil and paper, so there are several ways to get started in choosing a range. If the function is a polynomial, use the highest power to estimate the values at the endpoints of the interval you are using. For simple trigonometric functions with sin and cos functions examine the amplitude coefficient to estimate the amplitude of the oscillations. In general, the idea is to get some reasonable bounds for the range of the function. Then, the **ZFIT** or *zoom to fit* command on the TI-86 calculator can be used to choose more accurate values for **yMin** and **yMax**.

Once we have a first guess at the dimensions along the y axis, select **ZOOM**, press MORE and select **ZFIT**. The **ZFIT** command calculates the minimum and maximum values of the function on the closed interval [**xMin,xMax**], and sets these to be the new values for **yMin** and **yMax,** respectively. Then, a new graph is drawn with these new window settings.

You might wonder about functions that do not have a maximum or minimum on an interval. For example, the function $f(x) = 1/(1 - x^2)$ on $[-5, 5]$

has asymptotes at $x = \pm 1$. This problem is not encountered since the function is represented on the calculator as a table of data points and the problem of finding the maximum or minimum from a table is always a well defined problem. You may find that for this kind of function, **ZFIT** does modify your window parameters but there will not be major changes in their values. In this case, **BOX** might be a better choice for zooming.

A common mistake which occurs when choosing a window for graphing a trigonometric function can be demonstrated by considering

$$g(x) = \cos(20\pi x)$$

The cosine and sine functions oscillate periodically between -1 and 1. But the following example plots seem to indicate otherwise. The first plot has window dimensions of $[-3, 3] \times [-3, 3]$

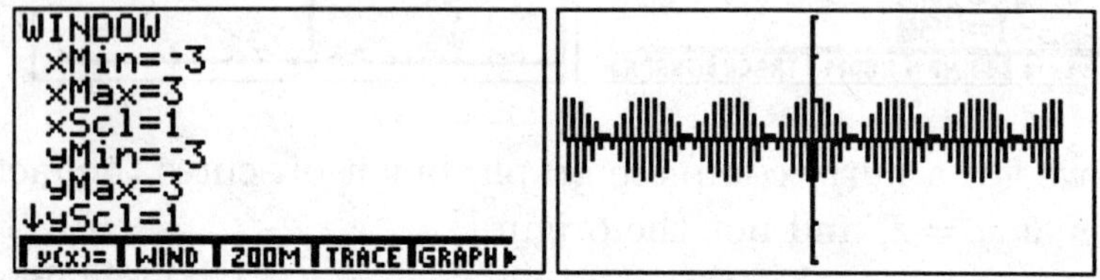

and the second plot has window dimensions $[-6.3, 6.3] \times [-3.1, 3.1]$.

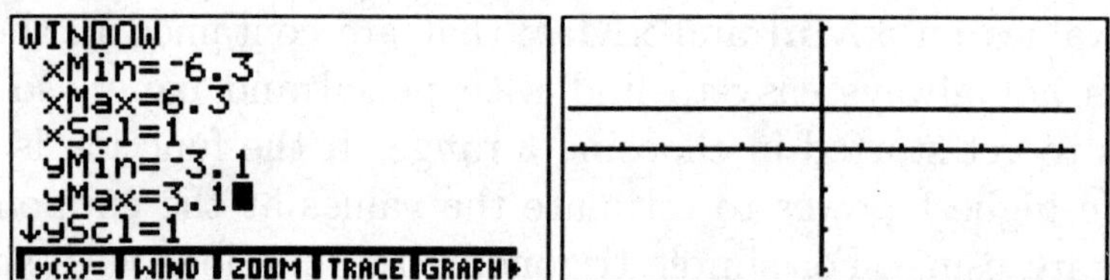

Both graphs are clearly wrong and require some explanation. They can be understood by noting that points are only plotted at one of the pixels which are spaced evenly by

$$\Delta x = \frac{xMax - xMin}{126}$$

In the second case the pixels are 0.1 apart and $\cos(20\pi \cdot 0.1n) = \cos(2n\pi) = 1$ for n an integer. Hence the graph is shown as a horizontal line. A similar analysis can be done for the first incorrect graph. A correct graph can be found by noting the period of this trig function. The period T of $\sin \omega t$ or $\cos \omega t$ is $T = 2\pi/\omega$. In this example the period is 0.1, so a window should be chosen so that only a few oscillations are shown. For example set the window dimensions to be $[-0.2, 0.2] \times [-1.5, 1.5]$.

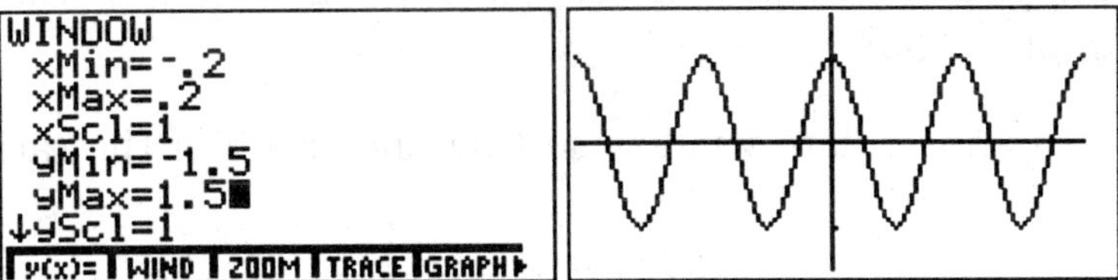

It's no wonder the two previous graphs are so wrong, since the windows chosen require 60 and 126 oscillations to be drawn on a screen that is just over two inches wide!

Our final example shows another consequence of replacing a continuous variable with a table of values. The graph of $h(x) = \sqrt{10 - x^2}$ is a semi-circle of radius 10. This function has the domain $[-\sqrt{10}, \sqrt{10}]$. The following screen shot shows the graph of $f(x)$ with window dimensions $[-4, 4] \times [0, 4]$.

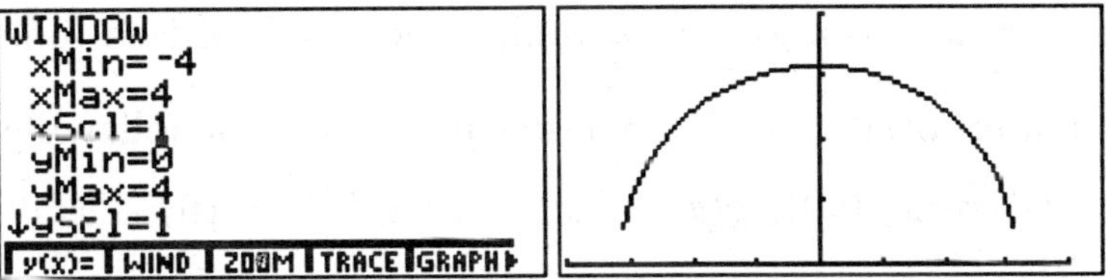

Notice that part of the graph is missing near the endpoints. Recall that functions are only plotted at a finite number of points spaced evenly over the interval. It is almost always the case that the endpoint of the domain will not correspond to one of the pixels where a point will be plotted. If we use **TRACE** near the endpoint, we see exactly how the points $\pm\sqrt{10}$ are skipped over. This is why it is so important to know the graphs of the basic functions. Otherwise, we can be misled by the graphs produced by a graphing calculator.

Exercises

1. Plot $f(x) = e^{-1000x} + x$ on a window of size $[0, 1] \times [-1, 2]$. Does this window show all the features of the function. If not, what is a better window?

2. What is the domain of the function $g(x) = x^{4/3}$? Plot the graph of this function in an appropriate window. Do you get a result consistent

with the domain. If not, re-enter the function in a new way so that the correct graph is plotted. Compare the functions $g(x)$ and $h(x) = |x|^{4/3}$. What happens if you enter a decimal approximation for the exponent, for example 1.333?

3. Choose an appropriate window so that the graph of the rational function

$$R(x) = \frac{x + 4}{x^2 + x - 6}$$

shows the features of its asymptotes, but does not show any extraneous lines where the asymptotes occur. Evaluate the function at $x = 0$ to find the y-intercept.

4. An amplitude modulated signal is modeled by the function

$$S(t) = \cos\left(\frac{t}{3}\right)\cos(12t)$$

Use one or more windows to examine the graph of this function.

5. Find an appropriate viewing window for each of the following functions:

$f(x) = 50\cos(x/100)$, $g(x) = x^4 - x^2$, $h(x) = (64 - x^2)^{1/4}$, $k(x) = \ln|x^2 - 9|$

2.4　Parametric Equations

A function given parametrically by

$$\begin{aligned} x &= f(t) \\ y &= g(t) \end{aligned}$$

with $a \le t \le b$ can be graphed on the TI-86 by first setting the type of graph in the **MODE** options to **Param**. Press $\boxed{\textbf{2nd}}$ [**MODE**], highlight **Param** and press $\boxed{\textbf{ENTER}}$.

If we press $\boxed{\textbf{GRAPH}}$, we see the choice $\textbf{E(t)}=$ instead of $\textbf{y(x)}=$. Now select $\textbf{E(t)}=$ and notice the appearance of the $\textbf{E(t)}=$ input screen.

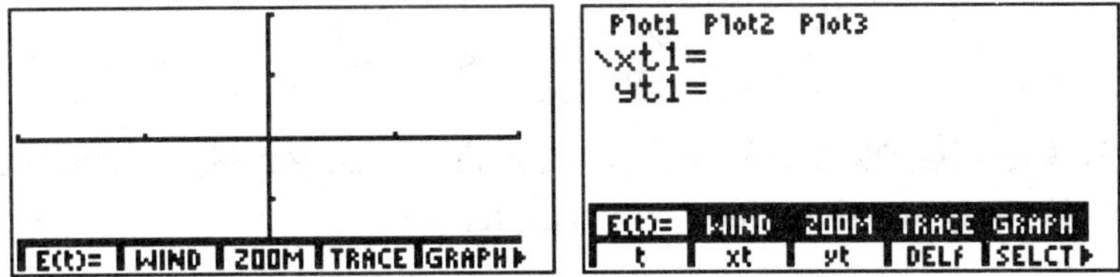

For convenience, the variable t is now one of the function key choices. For example, consider the curve given parametrically by

$$x = \cos(t), \quad y = \sin(2t)$$

We input these expressions for $\textbf{xt1}$ and $\textbf{yt1}$ below.

Now we just need to set the interval for t and the window dimensions. Select $\textbf{WIND}$. As above, the screen looks a little different. There are three additional numbers to enter. Besides the usual ones for the dimensions in the x and y directions, there are $\textbf{tMin, tMax}$ and $\textbf{tStep}$, which are used to set the range of values of the parameter t and the size of the step between points plotted. Note that for a given setting of $\textbf{tStep}$, the number of points plotted is

$$\frac{tMax - tMin}{tStep} + 1$$

In general, it is best to choose a value of $\textbf{tStep}$ so that this quantity is at least 50. Our choices are shown along with the graph in the next three screen shots.

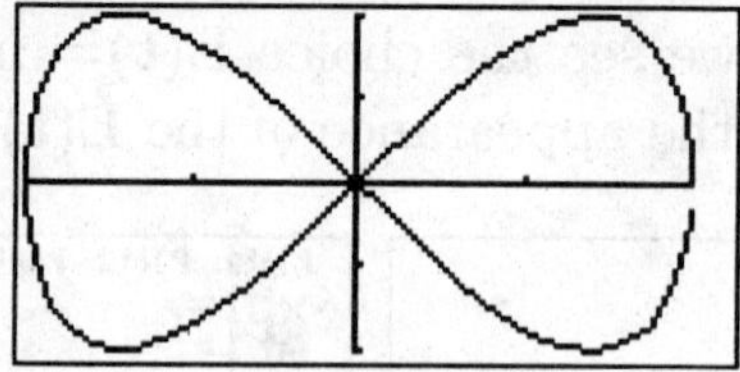

When Is a Circle Not a Circle? Let's consider the circle of radius 1, centered at the origin, given parametrically by

$$x = \cos t$$
$$y = \sin t$$

with $0 \le t < 2\pi$. If we choose the window $[-2, 2]$ by $[-2, 2]$ for a radius of 1, we get the following display.

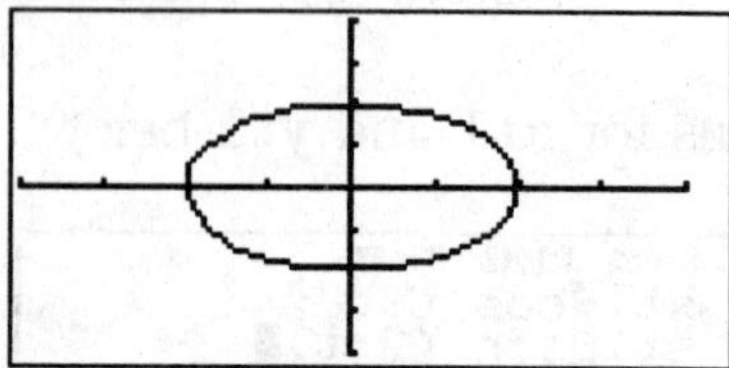

Our circle looks more like an ellipse than a circle! This is due to the dimension of the screen being longer along the x-axis than along the y-axis. The remedy for this and similar situations is to use window dimensions that are roughly twice as wide as they are tall. Consequently, a better choice would be $[-2, 2] \times [-1, 1]$.

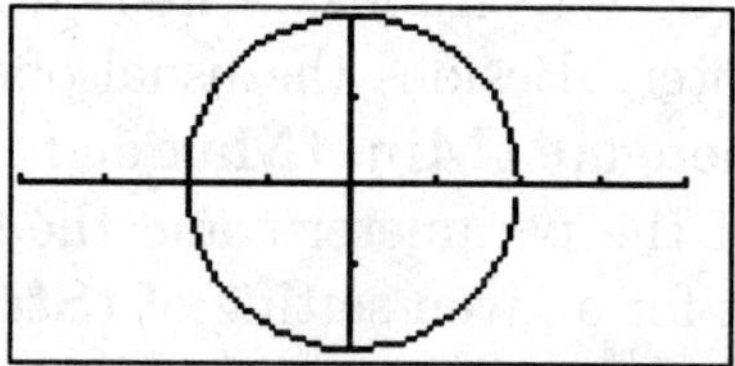

This is also a good place to demonstrate that sometimes a bad choice of **tStep** can be good. Since the TI-86 actually plots points along the circle and connects the dots, the circle shown above is really a regular polygon. The number of sides is determined by the size of **tStep**. The plots corresponding to the choices **tStep**$=2\pi/3$ and **tStep**$=2\pi/5$ are shown below.

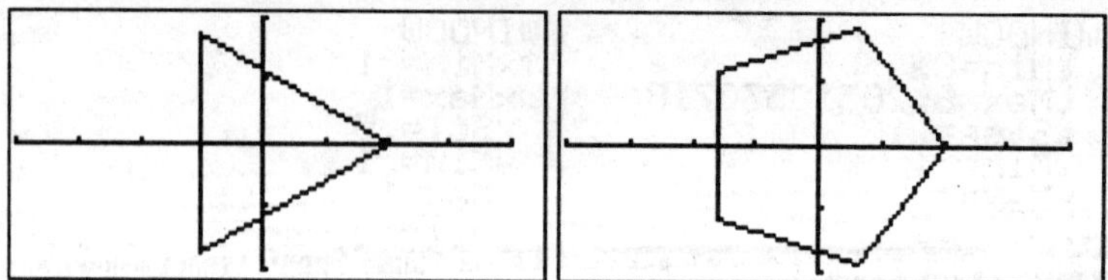

Exercises

1. Plot the ellipse given by

$$x(t) = 3\cos(t), \quad y(t) = 4\sin(t)$$

   for $0 \le t \le 2\pi$.

2. Plot the graph of the parametric function

$$x(t) = 5\cos(t) + \cos(5t), \quad y(t) = 5\sin(t) - \sin(5t)$$

   on the interval $0 \le t \le 2\pi$. This curve is called a hypocycloid. Evaluate the function at $t = \pi/2$.

3. Plot the parametric equations

$$x(t) = \sin(3t), \quad y(t) = \cos(4t)$$

   Choose an interval for t so that the complete picture is drawn.

4. Parametric equations for the path of a projectile with initial position $(0,0)$ are given by

$$x(t) = \frac{v_0 \cos(\theta_0)}{k}\left(1 - e^{-kt}\right)$$

$$y(t) = \frac{1}{k}\left[\left(v_0 \sin(\theta_0) + 32/k\right)\left(1 - e^{-kt}\right) - 32t\right]$$

   where k is the drag coefficient (due to air resistance) divided by the mass of the projectile, v_0 is the speed of the projectile at time $t = 0$, θ_0 is the angle between the trajectory and a horizontal line at time $t = 0$, $x(t)$ is the horizontal displacement at time t, and $y(t)$ is the vertical height of the projectile at time t. Note that at some time t the projectile will hit the ground. Suppose $v_0 = 100\frac{ft}{sec}$, $\theta_0 = \pi/4$ radians, and $k = 0.10$. Graph the trajectory of the projectile and find the time t at which it strikes the ground.

5. Return to the previous exercise. Experiment with different values of θ_0 to obtain an estimate for θ_0 which maximizes $x(t_{ground})$ where t_{ground} is the time at which the projectile strikes the ground.

2.5   Inverse Functions

The basic building blocks for functions of calculus have already been programmed into the TI-86 along with their inverses. For convenience the basic function is the key and the inverse function is the $\boxed{\textbf{2nd}}$ key. When there is more than one way to define an inverse function, the standard is used. For example, the range of $\cos^{-1} x$ is $[0, 2\pi]$.

Unfortunately, when basic functions are combined to form a new function, it is no longer a simple matter to determine the inverse of the function. The first step is to determine whether a given function has an inverse, and if it does, to plot its graph. This is easily done with the TI-86. For example, suppose

$$f(x) = x^3 + \frac{1}{100}x + 1, \quad \text{for} \ -2 \le x \le 4$$

To determine whether $f(x)$ has an inverse, we graph $f(x)$ in the usual way. First make sure **Func** is selected in the [**MODE**] window, and set the window dimensions.

Finally, input $f(x)$ as **y1** and create the graph.

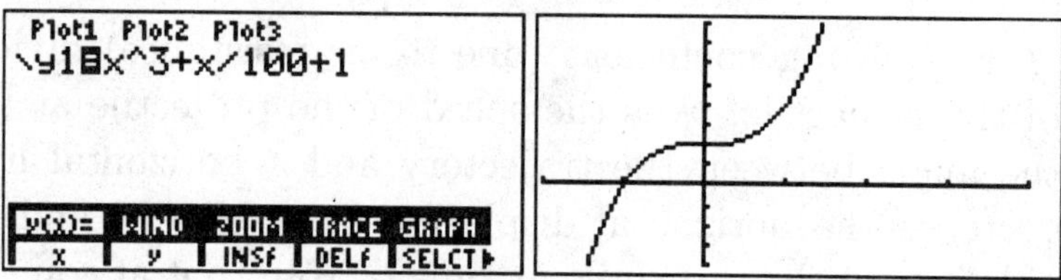

The graph seems to satisfy the horizontal line test. Consequently, the function is invertible. Although we can not easily find a formula for the inverse of $f(x)$, we can use the **DrInv** command to sketch the inverse. This command can be found by either pressing $\boxed{\textbf{MORE}}$, selecting [**DRAW**] and pressing $\boxed{\textbf{MORE}}$ three more times, or by pressing $\boxed{\textbf{2nd}}$ [**CTLG-VARS**], selecting **CATLG** and scrolling. Either method pastes the **DrInv** command to the home screen. Complete the command as shown to graph the inverse of $f(x)$.

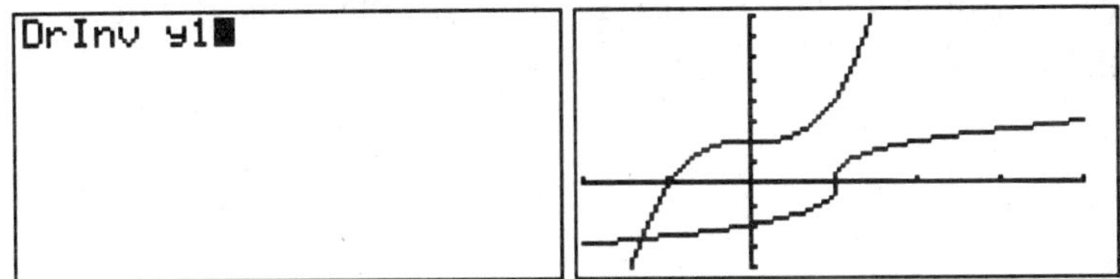

Of course this plot doesn't look quite right. The reason is simple. The window dimensions are not square. Reset the window dimensions by selecting **ZSQR** from within the **ZOOM** menu. Then re-execute the command line **DrInv y1** from the home screen.

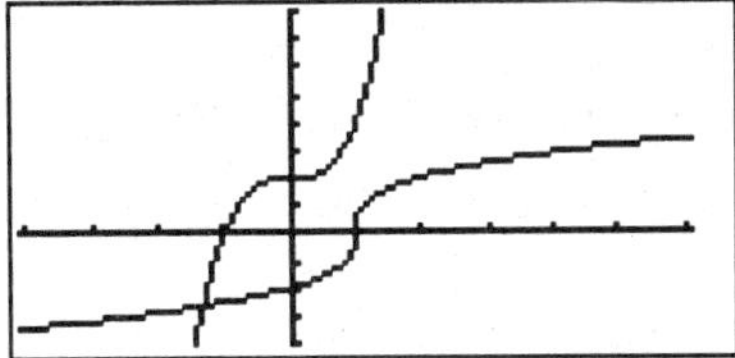

Now the graphs look symmetric across the line $y = x$.

Note that even if $f(x)$ doesn't have an inverse, the **DrInv** command will draw the graph of the reflection about $y = x$. It is up to us to decide whether the graph has significance in terms of inverse functions.

Since no equation has been given for the inverse graphed above, it is not possible to use the trace or the evaluation commands to identify points on the graph of the inverse, $f^{-1}(x)$. If you want to zoom in on a particular feature of this graph the result will be that only $f(x)$ will be displayed. So you will have to redraw the inverse function on this new window.

There are two ways to handle the problem of evaluating $f^{-1}(x)$ at a particular value of x (without having a formula for the inverse). One approach is to use the **SOLVER**. Since $y = f^{-1}(x)$ is the solution to $x = f(y)$, enter

$$y^3 + y = x$$

in the solver screen, choose a value of x, and then calculate y.

The other approach is to plot the inverse using a parametric representation where the trace function can be used. Any function $y = f(x)$ can be represented parametrically by

$$
\begin{aligned}
x &= t \\
y &= f(t)
\end{aligned}
$$

and since the inverse interchanges the coordinates, the inverse $f^{-1}(x)$ is given parametrically by

$$
\begin{aligned}
x &= f(t) \\
y &= t
\end{aligned}
$$

In this format it is easy to use the **ZOOM**, **TRACE** and **EVAL** commands to find details about the inverse function. Try this!

Exercises

1. Plot the function $f(x) = 2^x$ for $-4 \leq x \leq 4$. Does this function have an inverse? Explain. If so, plot the inverse. Also, interpret the inverse function in terms of logarithms.

2. Determine the domain of the function $f(x) = \sqrt{x^2 - x}$. Does this function have an inverse? If so, then plot the inverse and give a formula for the inverse.

3. Graph the equations $y = x^2 - 1$ and $x = y^2 - 1$ simultaneously. How many distinct regions in the plane are bounded by these two graphs?

4. Graph the equations $y = \sin(x)$ and $x = \sin(y)$ simultaneously in the window $[-2\pi, 2\pi] \times [-2\pi, 2\pi]$. Repeat with $y = \cos(x)$ and $x = \cos(y)$.

Chapter 3

Limits and Derivatives

In this chapter we explore how the TI-86 calculator can be used to find limits and derivatives. Complementary material can be found in Chapter 3 of Stewart's **Calculus**.

3.1   Limit of a Function

A combination of graphs and function values can be used to investigate limits. While this type of investigation is not rigorous, it can give a strong indication of the limiting value. We apply this approach to some limits below. In addition, we indicate some pitfalls associated with this method.

Let's start by considering the limit

$$\lim_{x \to 0} \frac{\sin x}{x}$$

To use the graphing approach enter $\sin x/x$ as the function **y1**, set the window dimensions to be $[-4, 4] \times [\ 1.5, 1.5]$ and graph **y1**.

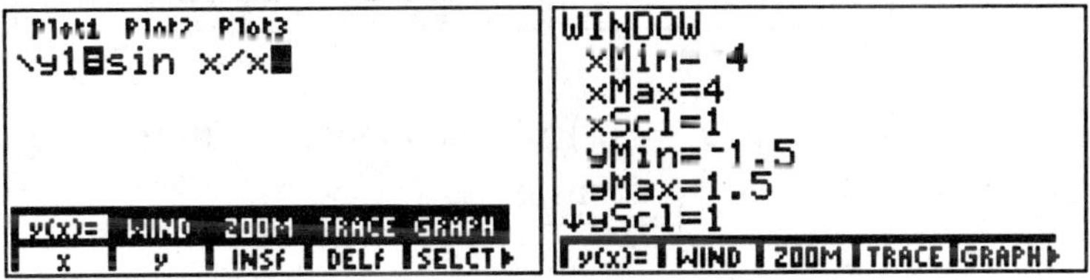

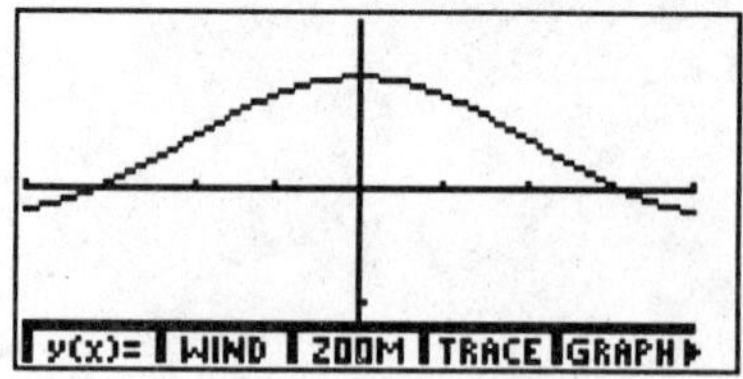

The graph indicates that the limit exists at $x = 0$. We can use **EVAL** (press $\boxed{\textbf{MORE}}$ $\boxed{\textbf{MORE}}$ $\boxed{\textbf{F1}}$)to evaluate $f(x)$ for x close to 0.

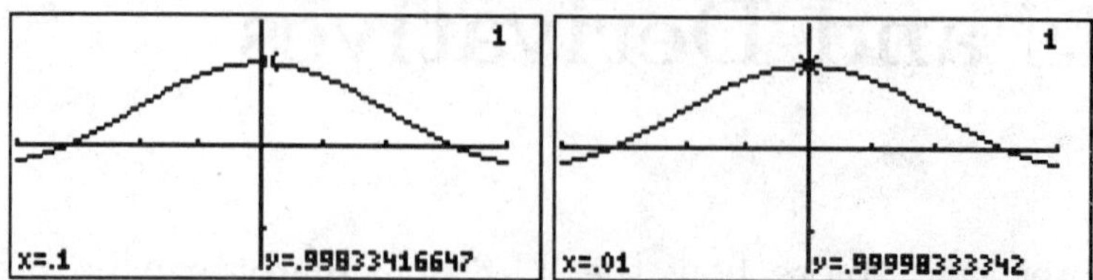

The values above indicate that the limit is 1. An even stronger case is made by examining the graph of the function on a very small interval, for example, $[-0.1, 0.1]$.

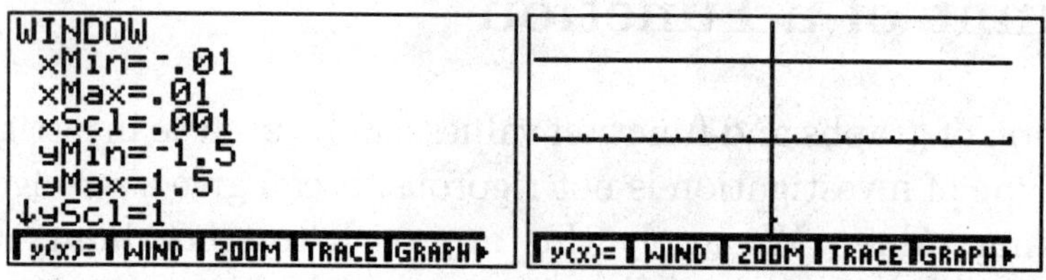

On this interval, the graph appears to be the horizontal line $y = 1$. This seems to confirm that the limit should be 1. If we check the **TABLE** then we obtain more reinforcement.

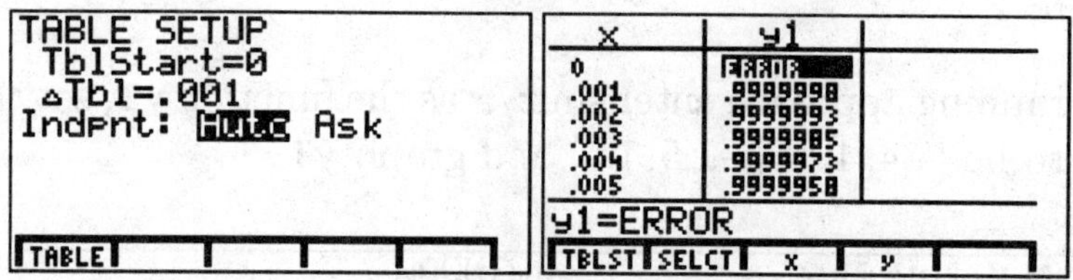

(Notice the error message at $x = 0$.) Consequently, our investigation leads us to conclude that

$$\lim_{x \to 0} \frac{\sin x}{x} = 1$$

In fact, this can be proved rigorously using the squeeze theorem.

NOTE: This same kind of numerical procedure without graphs or tables can be done using the **evalF** command from the home screen.

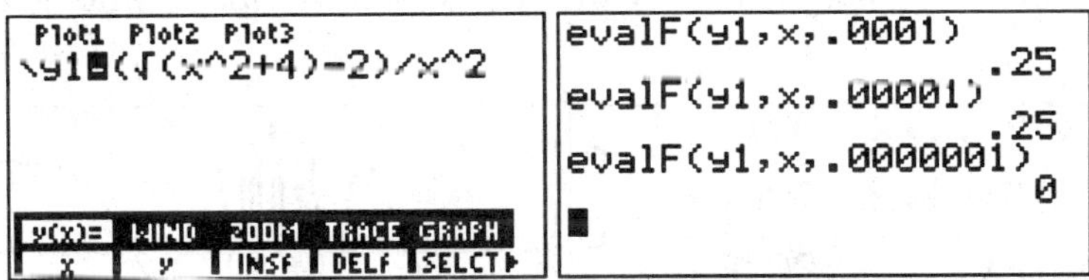

(This is a great place to use $\boxed{\textbf{2nd}}$ [ENTRY].)

The case of one-sided limits is handled in an identical fashion with the only difference being the choice of evaluation points. For a limit that approaches from above ($x \to a^-$) only choose values of x that are smaller than the value of interest. The opposite is the case for a limit where $x \to a^+$. The limits of piecewise defined functions are handled just as easily once the function is defined in the graphing screen as described in the previous chapter.

Sometimes roundoff error can give erroneous conclusions about a limit. For example, consider the limit

$$\lim_{x \to 0} \frac{\sqrt{x^2 + 4} - 2}{x^2}$$

It is not hard to show that this limit is $\frac{1}{4}$. However, some strange things happen if we evaluate the function at smaller and smaller values of x. At first, we see values close to the correct value. But near 10^{-6}, roundoff error begins to cause trouble. In fact, it appears (erroneously) from the values below that the limit is 0.

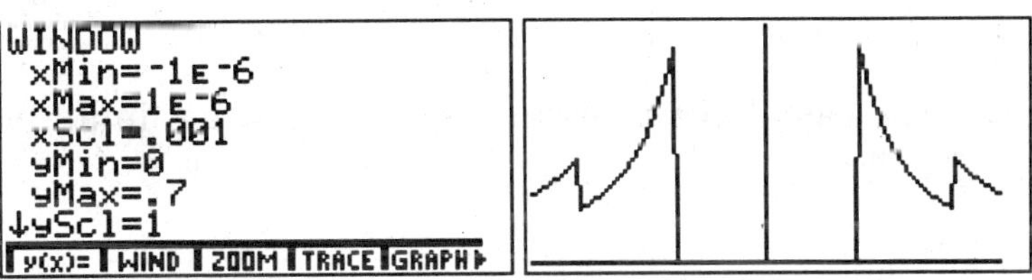

Further evidence of the trouble that roundoff causes can be seen by graphing **y1** in a $[-10^{-6}, 10^{-6}] \times [0, 0.7]$ window.

A few words of warning are in order. First, when using a calculator to obtain information about limiting values of a function $f(x)$ as x approaches a, be careful in choosing the values of x. Always choose values of x on both sides of the value of a in a two sided limit. A function can have distinct behaviors on each side. For example, the function $f(x) = |x|/x$ has this property. If this function is input as **y1**, and the **evalF** command is used to evaluate the function for values of x to the right of 0, then it is easy to (*mess up*) by concluding that $f(x)$ approaches 1 as x approaches 0. However, the graph shown below clearly indicates this is false.

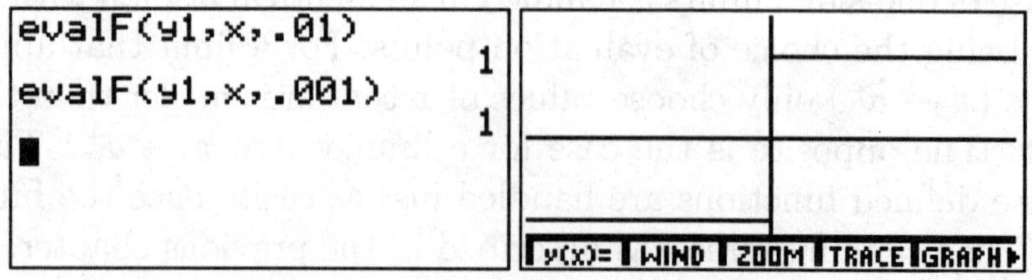

Second, it is not always a good idea to stick with a fixed pattern of values of x (like those illustrated in the last example). It is better to inject some randomness into your choice of x values. The limit $\lim_{x \to 0} \sin \frac{\pi}{x}$ illustrates this point nicely. If you choose $x = 10^{-n}$ where n is an integer, the value of

$$\sin \frac{\pi}{x} = \sin(\pi/10^{-n}) = 0$$

for all values of n. One might like to conclude that the limit is zero based on this numerical evidence, but the graph below indicates otherwise. Input the function $\sin(\pi/x)$ as **y1** to obtain the function values and graph shown below.

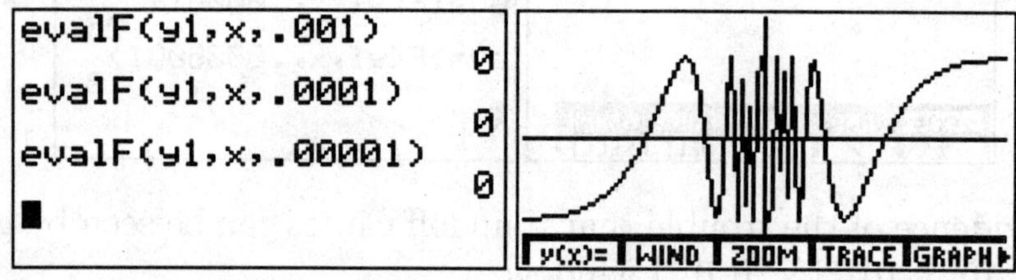

The truth is that no limit exists.

Exercises: Use the calculator to estimate the following limits both graphically and numerically.

1. $\lim_{x \to 2} \frac{x^3 - 8}{x - 2}$

2. $\lim_{x \to 0} \frac{e^x - 1}{x}$

3. $\lim_{x \to -2} \frac{x^2 - 4}{|x + 2|}$

4. $\lim_{x \to 0+} x \ln x$

5. $\lim_{x \to 0} x^2 \sin \left(\frac{1}{x} \right)$

6. $\lim_{x \to 0} \frac{1 - \cos 2x}{x}$

7. $\lim_{x \to 0} (1 + 2x)^{3/x}$

8. $\lim_{x \to 5} \left(\ln \left(\sin(x - 5) \right) + x - \ln(2x - 10) \right)$

3.2   Continuity

A simple graphical interpretation of continuity is that a function's graph has no jumps, gaps or breaks. Continuity is more precisely defined using the limit definition. That is,

$$\lim_{x \to a} f(x) = f(a).$$

How does the idea of continuity show up when graphing a function using the TI-86 calculator? This is not a simple question. Remember that we have to be careful in our claim that the graph drawn is completely correct. To illustrate this point, consider the simple piecewise function

$$f(x) = \begin{cases} -x & \text{if } x \leq 1 \\ x^2 & \text{if } x > 1 \end{cases}$$

which is defined with the rule $(x \leq 1) \, (-x) + (x > 1) \, x^2$.

We don't need a graphing calculator to plot this function. However, this simple function illustrates a point which can cause problems when more complicated functions are analyzed.

The plot of this function appears to be a continuous curve with no breaks or jumps.

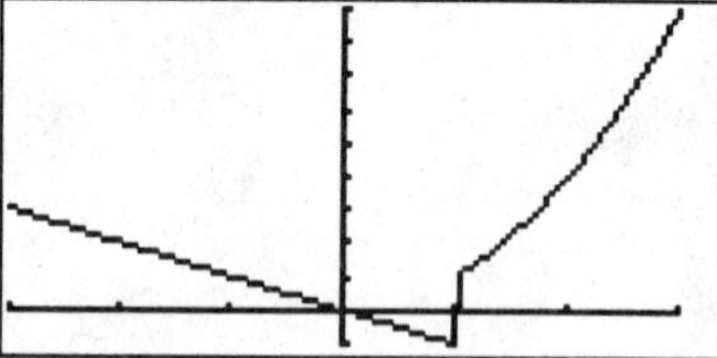

There is simply a very sharp rise in the plot near the transition from the line to the parabola. Even zooming in on this function doesn't produce a correct graph. This is because the calculator is plotting points and connecting them with line segments. You can either live with this or set the plot options to **DrawDot** instead of **DrawLine,** so that no connection is made.

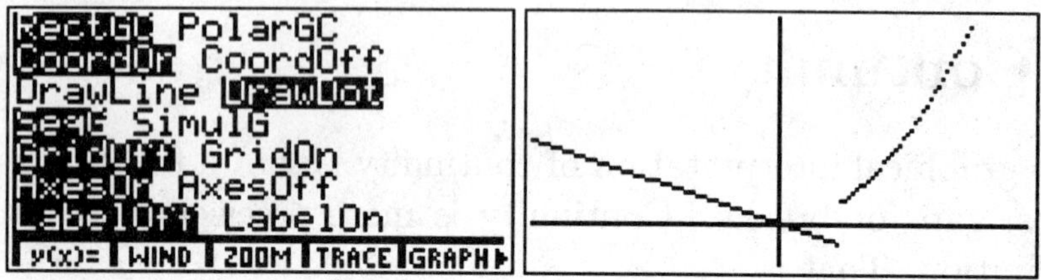

Although we still don't get a perfect picture of the function's graph, it appears as though the function has a jump discontinuity at $x = 1$.

Some types of discontinuities are said to be removable since they can be repaired by defining the function to match its limiting behavior. For an example, consider the piecewise defined function

$$g(x) = \begin{cases} \frac{x^2-1}{x-1} & \text{if } x \neq 1 \\ 3 & \text{if } x = 1 \end{cases}$$

which is not so easy to graph correctly. The best way to see the graph is to use the fact that $(x^2 - 1)/(x - 1) = x + 1$ when $x \neq 1$. Enter $(x \neq 1)(x + 1) + (x == 1)3$ as **y1** for $g(x)$. Also choose the **ZOOM** option **ZDECM** so that $x = 1$ corresponds to a pixel. The screen shots show the **DrawLine** format which gives an incorrect graph and the **DrawDot** format showing the correct graph with a single point sitting all by itself.

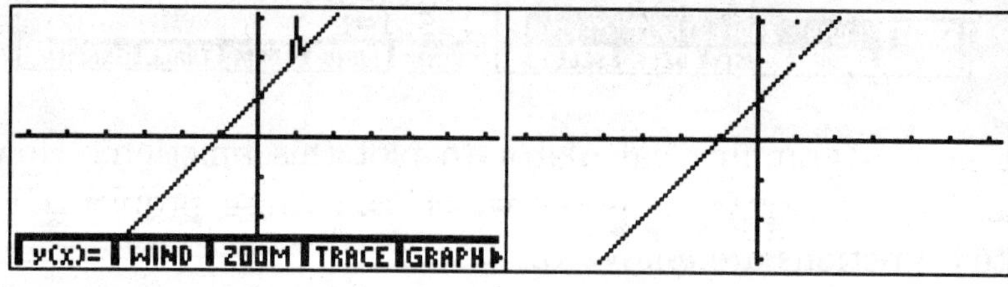

By evaluating the function near $x = 1$, we see that it is better to define $g(1) = 2$.

```
evalF(y1,x,.999)
                1.999
evalF(y1,x,1.001)
                2.001
evalF(y1,x,1)
                    3
■
```

This can also be seen by performing a simple computation:

$$\lim_{x \to 1} g(x) = \lim_{x \to 1} (x + 1) = 2$$

A more interesting problem associated with continuity can be illustrated by considering the function

$$f(x) = \begin{cases} x^2 + 2, & x < 1 \\ ax + b, & 1 \le x \le 3 \\ 3 - 2x + x^2, & x \ge 3 \end{cases}$$

Let's try to find values of a and b so that $f(x)$ is a continuous function. Note that $f(x)$ is given by a parabola, a line and a parabola on the intervals $(-\infty, 1)$, $[1, 3]$ and $[3, \infty)$ respectively. Consequently, we simply need to choose a and b so that $f(x)$ is continuous at $x = 1$ and $x = 3$. Now,

$$f(1) = a + b = \lim_{x \to 1^+} f(x)$$

and

$$f(3) = 3a + b = \lim_{x \to 3^-} f(x)$$

Also,

$$\lim_{x \to 1^-} f(x) = \lim_{x \to 1^-} \left(x^2 + 2\right) = 3$$

and

$$\lim_{x \to 3^+} f(x) = \lim_{x \to 3^+} \left(3 - 2x + x^2\right) = 6$$

Therefore, we need to find a and b so that

$$\begin{pmatrix} a + b = 3 \\ 3a + b = 6 \end{pmatrix}$$

We can solve this system using the TI-86 by pressing $\boxed{\textbf{2nd}}$ [SIMULT], entering the value 2 and pressing $\boxed{\textbf{ENTER}}$.

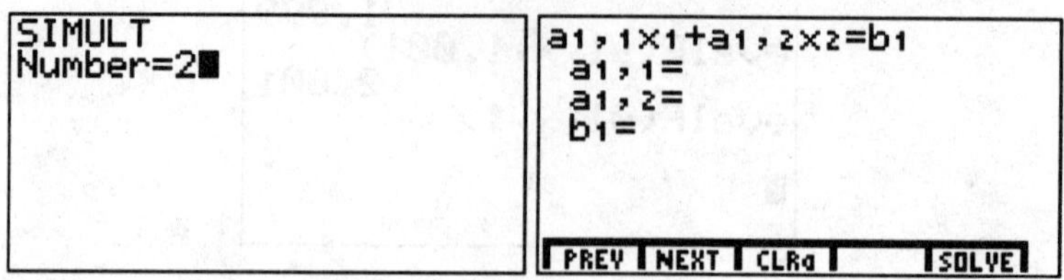

Then simply enter the coefficients for the equations. Press $\boxed{\textbf{ENTER}}$ after entering the first the coefficients for the first equation to move to the second equation.

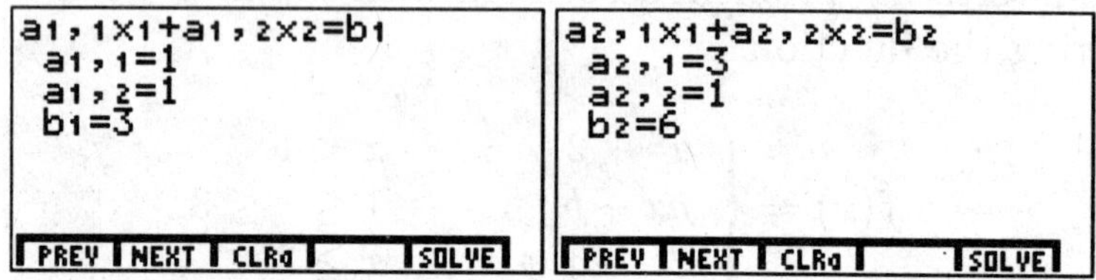

Finally, press $\boxed{\textbf{F5}}$ (i.e. **SOLVE**) to solve the system.

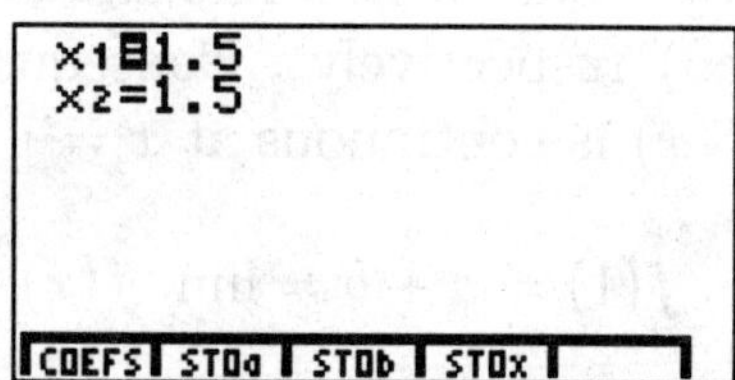

As a result, if we choose $a = 1.5$ and $b = 1.5$ then $f(x)$ is a continuous function.

Exercises

1. Find the values of a and b so that

$$f(x) = \begin{cases} 2x - 1, & x \leq 1 \\ ax^2 - bx, & 1 < x < 2 \\ 2x^3 - 8, & x \geq 2 \end{cases}$$

   is continuous.

2. Find values of a, b and c so that the function

$$g(x) = \begin{cases} 2x - x^2, & x \leq -1 \\ ax^2 + bx + c, & -1 < x < 3 \\ (x^2 - 9) / (x - 3), & x \geq 3 \end{cases}$$

is continuous and satisfies $g(2) = 3$.

3. Demonstrate graphically and numerically that

$$f(x) = \frac{\sin\left(\sin\left(\sin(2x)\right)\right)}{3\sin(x)}$$

has a removable discontinuity at $x = 0$. How should $f(0)$ be defined?

3.3   Limits with Infinity

The behavior of a function's graph near a vertical or horizontal asymptote can be numerically determined in much the same way as is done for other types of limits.

Recall that a vertical asymptote occurs for rational functions at the points where the denominator is zero. For the simple rational function

$$f(x) = \frac{1}{x - 1}$$

there is a vertical asymptote at $x = 1$. To investigate the behavior of $f(x)$ near $x = 1$, it is best to evaluate this function for values of x close to 1. As we saw in the last chapter, a graph with a vertical asymptote does not show up very well unless you choose an appropriate window. In fact, in some cases, any choice of window does not give an accurate representation of the graph. The graph and function values shown below indicate that $f(x)$ takes on larger and larger values as $x \to 1^+$.

This suggests that the function becomes unbounded for $x \to 1^+$; that is

$$\lim_{x \to 1^+} f(x) = \infty$$

A similar analysis for $x \to 1^-$ suggests that $f(x)$ becomes unbounded but negative.

Note that your calculator has an overflow of $\pm 10^{999}$ and will give an error message when an evaluation yields an answer larger than this. For most typical problems encountered in calculus, the overflow probably will not occur.

For any function with asymptotes, first carefully graph the function with strict attention paid to the window dimensions so that you can see the asymptotes. This may require looking at several windows to see more than one asymptote. For example, consider the graph of $g(x) = 1/(x^2 - 5)$. **This example is easy to analyze without the help of the calculator, but the process shown below can be applied to more complicated expressions.** Start by inputting the expression for $g(x)$ as **y1**.

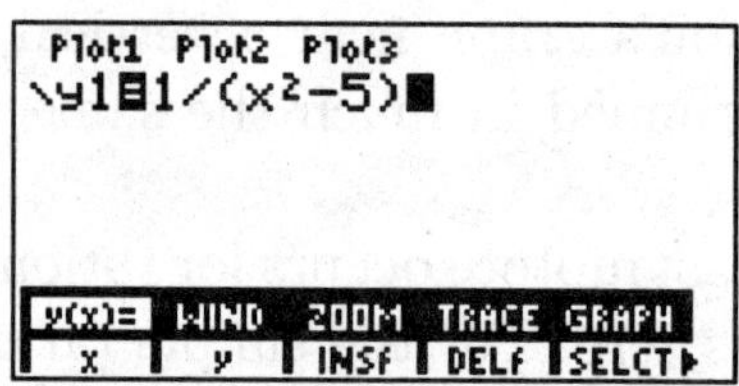

This function is plotted below in a $[-4, 4] \times [-4, 4]$ window.

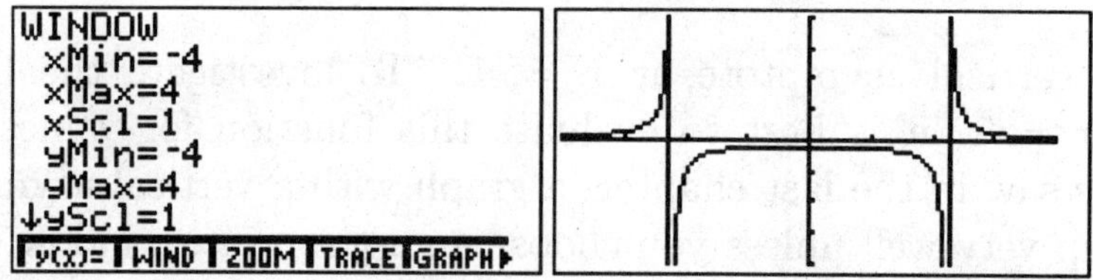

TRACE can be used to help identify the location of the vertical asymptotes.

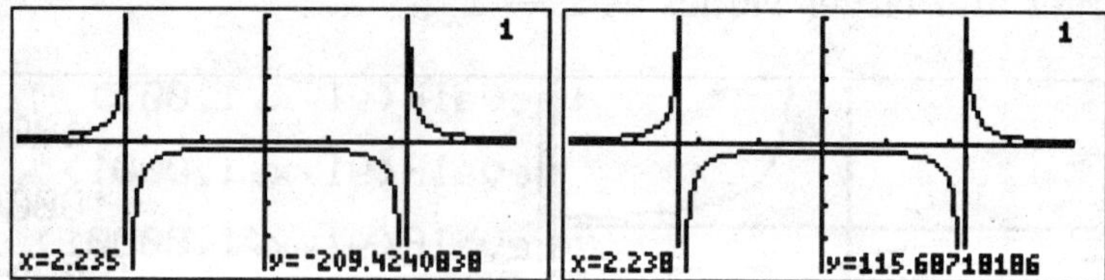

The **SOLVER** can also be helpful in locating where the asymptotes are located. The first screen below shows the result from the **SOLVER**. Note that it only returns one of the roots. The second screen shows the result of substituting values close to the solution from the **SOLVER** into the function.

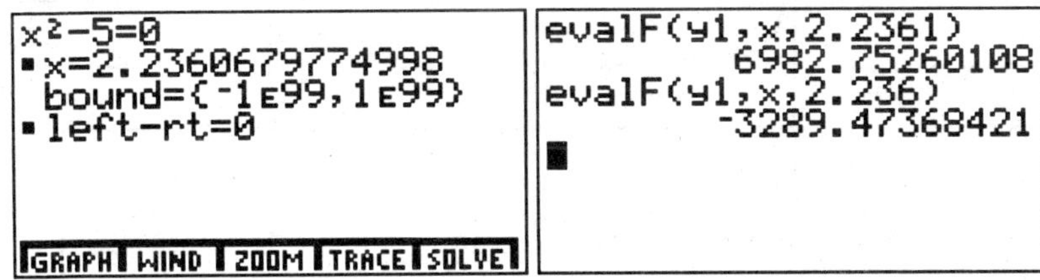

Actually, since the denominator $x^2 - 5$ is a polynomial, we can find all of its roots by using $\boxed{\text{2nd}}$ [**POLY**], inputting the value 2 and pressing $\boxed{\text{ENTER}}$.

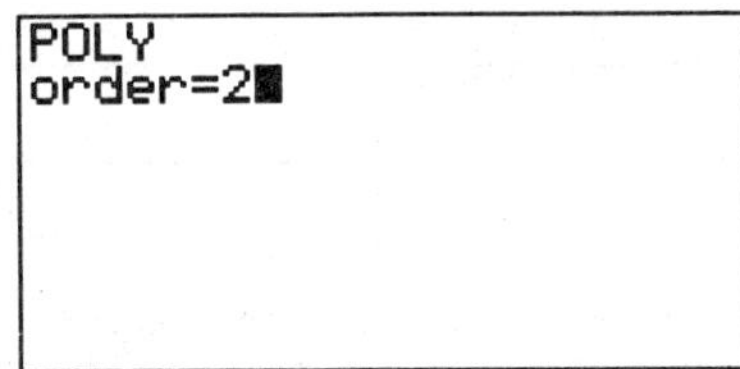

Then input the values shown in the first screen below and press $\boxed{\text{F5}}$ (i.e. **SOLVE**).

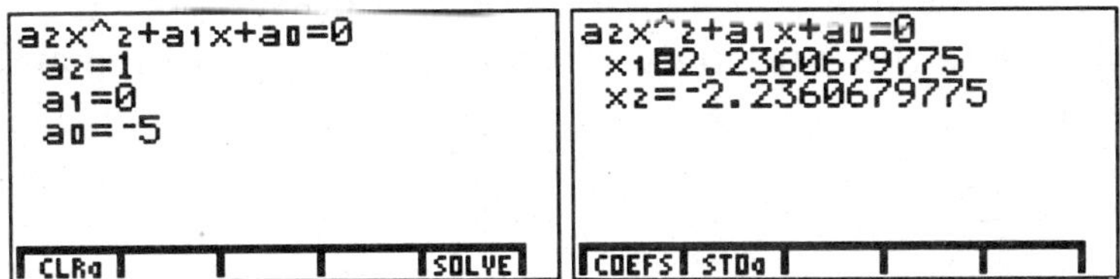

The behavior of functions near horizontal asymptotes is found by numerically evaluating a function for large positive or negative values of x. For example, to evaluate the limit

$$\lim_{x \to \infty} \frac{x+1}{x-1},$$

enter the function $(x+1)/(x-1)$ for **y1**. The graph should be viewed from a distance to see the connection between the limiting value as $x \to \infty$ and the horizontal asymptote. Choose a window which covers a large range of x values. Use either the **EVAL** from the graphing screen, the **evalF** command or **TRACE** to evaluate the function for larger and larger values of x. You will find the function values will not change much for large x due to the asymptote present. This suggests that the limiting value is 1.

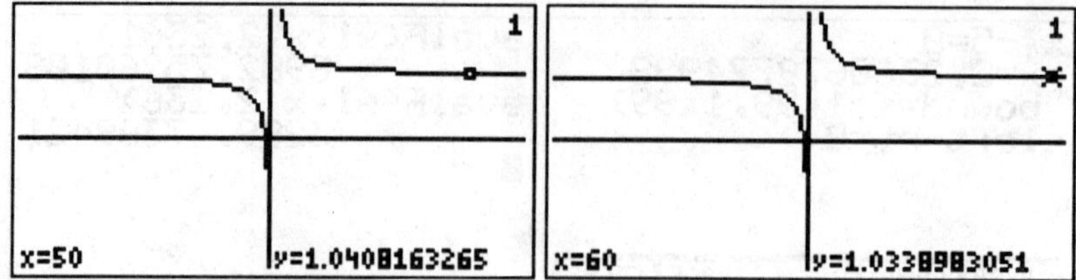

If you include the graph of the horizontal asymptote $y = 1$, then you have a nice picture of the large x behavior.

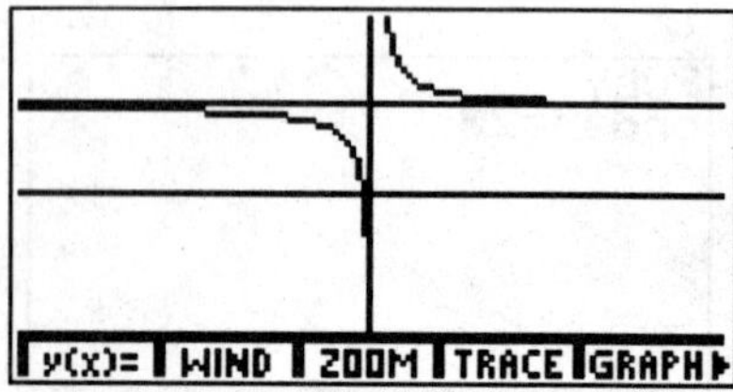

Exercises: Find the horizontal asymptote(s) for the functions in exercises 1-5. Then graph the function and the horizontal asymptote(s) on the same display to illustrate the function's behavior for large positive and large negative values of x.

1. $f(x) = \frac{\sqrt{x^2 - 4}}{x}$

2. $g(x) = x^{10} e^x$

3. $h(x) = \frac{x^3 + 7x^2 + 8x}{2x^3 + 5}$

4. $F(x) = \frac{3x^2 + 4x + 8}{x|x| + 9}$

5. $G(x) = e^{-x^2} \sin \pi x$

6. Find the horizontal and vertical asymptotes of the function

$$f(x) = \frac{2x - 3x^3}{2x^5 - 9x^4 + 11x^3 + 9x^2 - 11x - 1}$$

Determine the behavior near each of the vertical asymptotes.

3.4 Tangent Lines

To see how the calculator can be used to study secant lines and tangent lines, enter the function $f(x) = \sqrt{x}$ in the graphing screen with a window size of $[0, 1.26] \times [-0.5, 1.5]$.

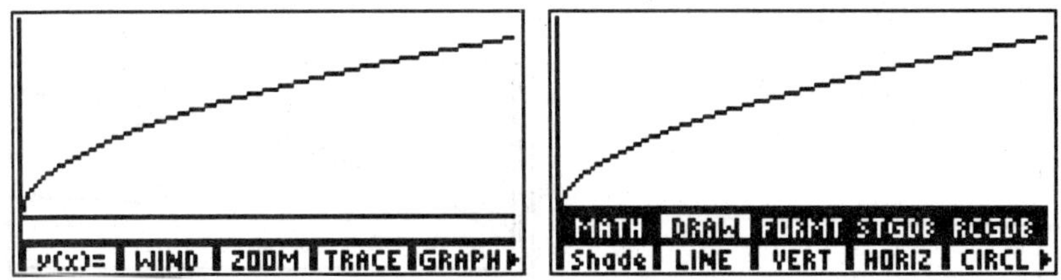

Once you have graphed the function, press $\boxed{\textbf{MORE}}$ once and then $\boxed{\textbf{DRAW}}$. The command **LINE** is started by pressing $\boxed{\textbf{F2}}$. Now you can draw any line interactively by using the arrow buttons to move the cursor to the location of one of the points that will be part of the line. The coordinates of the cursor are shown at the bottom of the display to help locate points. Press $\boxed{\textbf{ENTER}}$ to lock the location of the first point. Then use the arrow buttons to locate the second point and press $\boxed{\textbf{ENTER}}$ to complete the line segment.

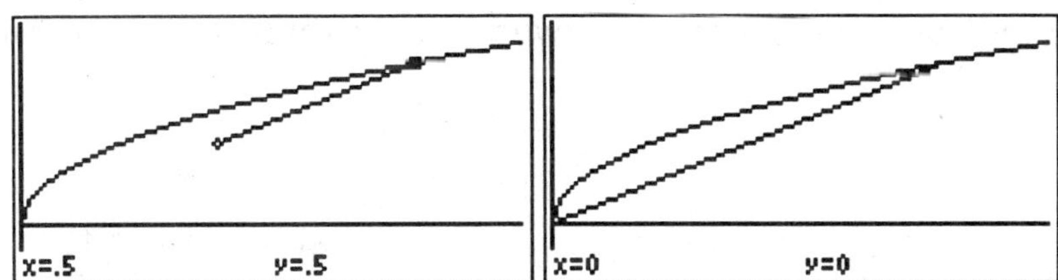

The screen shot shows the secant lines that connect $(0, 0)$ to $(1, 1)$ for the function $f(x) = \sqrt{x}$. A line segment can also be created by using the **Line** command from the home screen. Use the $\boxed{\textbf{ALPHA}}$ key, or press $\boxed{\textbf{2nd}}$ [CATLG-VARS] and select **CATLG** to find the **Line** command. Entering **Line(a,b,c,d)** will draw a line segment between the points with coordinates (a, b) and (c, d).

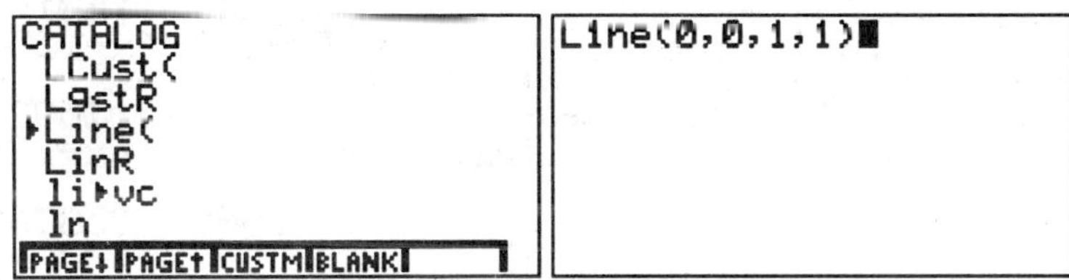

The line, $y = x + \frac{1}{4}$, drawn in the picture is the tangent line to this function at $x = \frac{1}{4}$. It can be drawn using the **TanLn** command (available from the **DRAW** menu). After activating this command, you are prompted with the command line **TanLn(** which you complete with the names of the function, say **y1**, followed by a comma and the x-coordinate of the point where you want to find the tangent line.

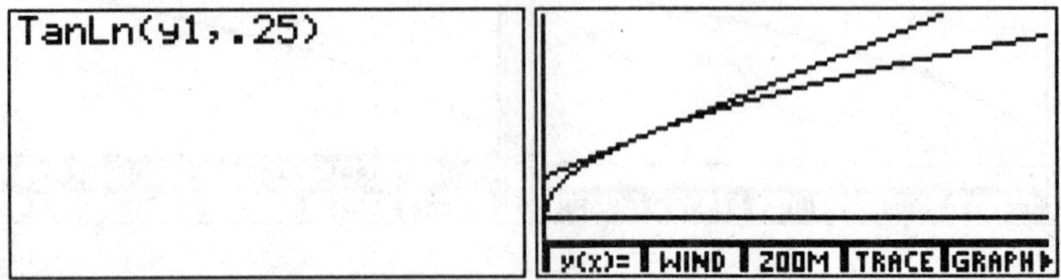

The **TanLn** command does not require any knowledge of the equation for the tangent line. If you happen to know the equation you can directly input it as a function instead.

Drawing secant and tangent lines to a curve can clutter up the screen very quickly. To delete all lines while keeping the function's graph, use the clear drawing command, **CLDRW**, from the **DRAW** menu.

The slope of a secant line that passes through two points on a function's graph, (x_0, y_0) and (x_1, y_1) can be investigated graphically and numerically. The slope is given by

$$m = \frac{y_1 - y_0}{x_1 - x_0}.$$

Consider the example function $f(x) = \sqrt{x}$ given above. Note that $f(1/4) = 1/2$ and $f(1) = 1$. Consequently, the slope of the secant line connecting $(\frac{1}{4}, \frac{1}{2})$ to $(1, 1)$ is given by

$$\frac{1 - 1/2}{1 - 1/4} = \frac{2}{3}$$

If we fix $(x_0, y_0) = (\frac{1}{4}, \frac{1}{2})$ and move the point (x_1, y_1) slowly towards (x_0, y_0), then the slopes of the corresponding secant lines are given by the calculations below.

(**2nd** [**ENTRY**] is very useful here). It appears that the slope is getting closer to 1 as x gets closer to $\frac{1}{4}$. Consequently, we might conclude that the slope of the tangent line to the graph of $f(x)$ at $x = \frac{1}{4}$ is 1.

The **TanLn** command computes the slope of the tangent line using the formula

$$m_T \cong \frac{f(x+\delta) - f(x-\delta)}{(x+\delta) - (x-\delta)},$$

i.e. by computing the slope of the secant line that connects the points $(x - \delta, f(x - \delta))$ and $(x + \delta, f(x + \delta))$. The number δ is a tolerance value that has a default setting of 0.001. This value can be changed by pressing **2nd** [**MEM**] and then pressing **F4** for **TOL**.

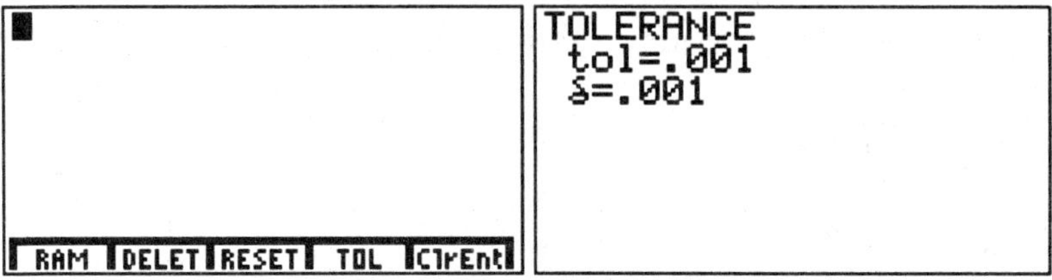

The default value gives fairly good accuracy for most situations.

Exercises

1. Plot the function $f(x) = \cos x$ on a window that shows a close in view of $x = 0$. Use the TI-86 to draw several secant lines that pass through the points $(0, 1)$ and $(a, \cos a)$ where a is chosen close to 0. You may want to zoom in on the origin to have a better view. What can you say about the limiting secant line as $a \to 0$? Use **TanLn** to draw the tangent line at $x = 0$.

2. Plot the function $g(x) = x^{1/3}$ on a window that shows a close in view of $x = 0$. Using the TI 86 to draw several secant lines that pass through the points $(0, 0)$ and $(a, a^{1/3})$ where a is chosen close to 0. You may want to zoom in on the origin to have a better view. What can you say about the limiting secant line as $a \to 0$? What happens if you use **TanLn** to draw the tangent line at $x = 0$?

3. Plot the function $f(x) = x^2 - 1$ in the window $[-2, 2] \times [-2, 3]$. Then plot its tangent lines at $x = -1$, 0 and 1.

4. Plot the function $f(x) = \sin x$ in the window $[-1, 7] \times [-1.5, 1.5]$. Then plot its tangent lines at $x = 0, \frac{\pi}{2}, \pi, \frac{3\pi}{2}$ and 2π.

5. Plot the function $f(x) = \sin\left(\frac{\pi}{2}x\right)$ in the window $[0, 4] \times [-1.5, 1.5]$ along with secant lines through $(1, 1)$ and each of the points $(2, 0)$, $(5/3, 1/2)$ and $(4/3, \sqrt{3}/2)$.

6. For each of the functions in parts a-d, compute the difference quotient

$$\frac{f(1+h) - f(1)}{h}$$

for $h = 0.1, 0.01, 0.001$. Give a guess for the slope of the tangent line at $x = 1$ based upon your calculations.

(a) $f(x) = 3x^2 - x + 2$

(b) $f(x) = 1/x$

(c) $f(x) = e^x$

(d) $f(x) = \sin^2(\pi x)$

7. Find, to two decimal places, the value of x in the interval $[0, 2]$ where the tangent line to the graph of $f(x) = x^3 - x^2 - x$ is parallel to the secant line through $(0, 0)$ and $(2, 2)$. Plot the graph of $f(x)$ and both lines in the window $[0, 2] \times [-2, 2]$.

3.5   Derivatives

The standard definition for the derivative is given by

$$f'(x) = \lim_{\delta \to 0} \frac{f(x + \delta) - f(x)}{\delta}$$

and an equivalent form is found by replacing δ with $-\delta$

$$f'(x) = \lim_{\delta \to 0} \frac{f(x) - f(x - \delta)}{\delta}$$

If each of these limits exist (i.e. $f'(x)$ exists), we can sum the expressions above and conclude that

$$\begin{aligned} f'(x) &= \lim_{\delta \to 0} \frac{f(x+\delta) - f(x-\delta)}{2\delta} \\ &= \lim_{\delta \to 0} \frac{f(x+\delta) - f(x-\delta)}{(x+\delta) - (x-\delta)}. \end{aligned}$$

This is just the statement that the derivative at x is the limit of the slope of the secant line that connects the points $(x-\delta, f(x-\delta))$ and $(x+\delta, f(x+\delta))$. If δ is quite small we would expect the derivative is approximated by the slope of such a secant line. This is the principle that allows the TI-86 calculator to numerically approximate derivatives. This approximation is called the *central difference approximation* to the derivative. The value of δ, called the tolerance, has a default value of 10^{-3}. This can be adjusted to a smaller value by editing its value from the 2nd [MEM] screen.

First derivatives can be calculated in two ways from the calculus menu 2nd [CALC].

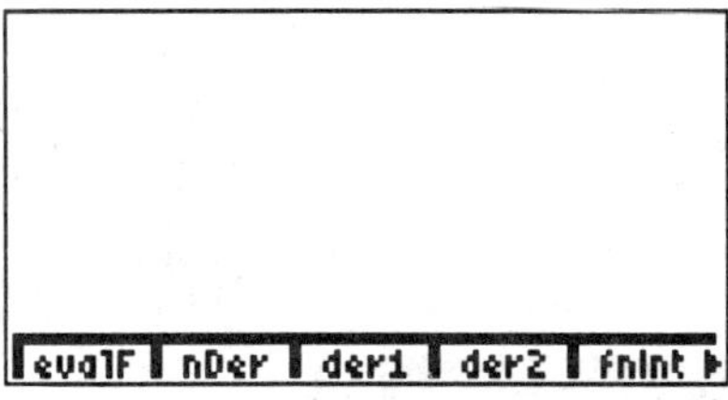

The two choices are, **nDer**, for numerical derivative, and **der1**, for first derivative. The **nDer** command is more general, and can used for all types of functions (including piecewise defined functions). The syntax for using this command is

$$\mathbf{nDer}(f(x), x, a)$$

where $f(x)$ is a function defined explicitly or implicitly using the **y1** notation. The second argument is the variable and the last is the number where the derivative is to be evaluated. It is susceptible to errors due to the numerical nature of the calculation. Usually a smaller value of δ can improve the results as seen in this example.

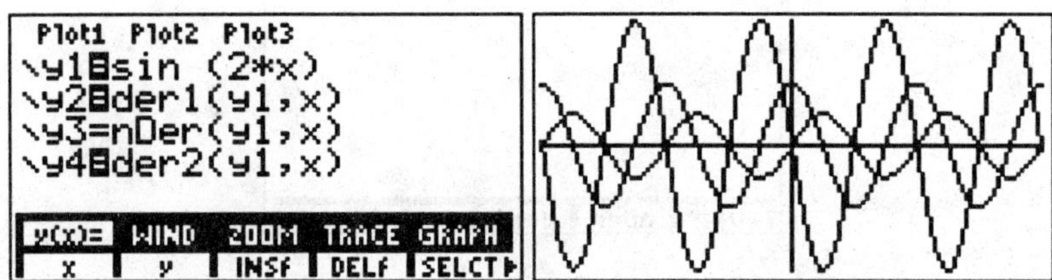

The **der1** command uses an algorithm that computes the derivative using the usual rules for differentiation. The syntax is the same for **der1** as it is for **nDer**. The resulting derivative values are extremely accurate. The disadvantage of this method is that it is limited to basic functions for which the calculator can apply the algorithm. This command cannot be used for evaluating derivatives of piecewise defined functions.

The second derivative can be computed by using the **der2** command. This command uses the same syntax as **nDer** and **der1**.

Since the derivative is itself a function, it can also be graphed. This can be done using either the **nDer** or **der1** commands. The function screen below shows how to enter the required functions in the graphing menu. Note that the third arguments of **nDer**, **der1** and **der2** are optional when graphing a derivative.

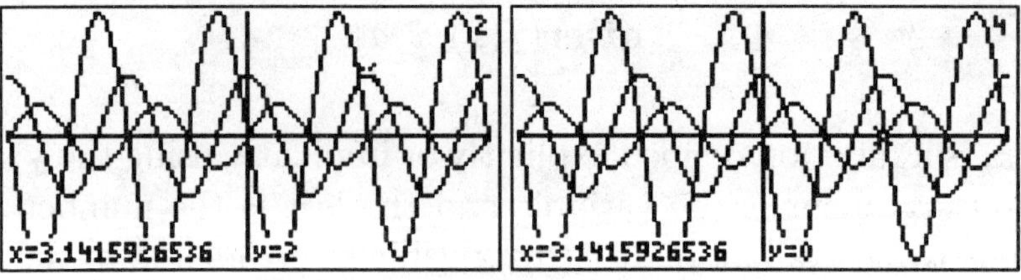

The graphing screen shows the function with its first two derivatives. The **TRACE** or **EVAL** commands can be used to evaluate the function or its derivatives at any point. The up and down arrow keys can be used to scroll through the function and derivative values.

Functions that have places where the derivative does not exist can also be handled. We simply need to watch for jump discontinuities which might not get drawn accurately. The example screen shot shows the absolute value function $g(x) = |x|$ and its derivative $g'(x)$, which has a jump discontinuity at $x = 0$. The absolute value function is the only function which is not smooth that the **der1** command handles correctly. Note that **TRACE** shows the derivative to be undefined at $x = 0$.

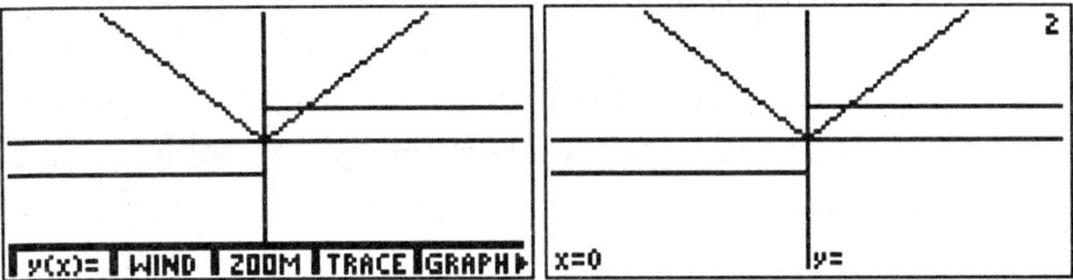

To plot derivatives of a more general piecewise function, be sure to use the **nDer** command, since **der1** will lead to the error message **INVALID**.

To graph the function

$$h(x) = \begin{cases} -x & \text{if } x < 1 \\ x^2 & \text{if } x \geq 1 \end{cases}$$

correctly, use the **nDer** command as shown below.

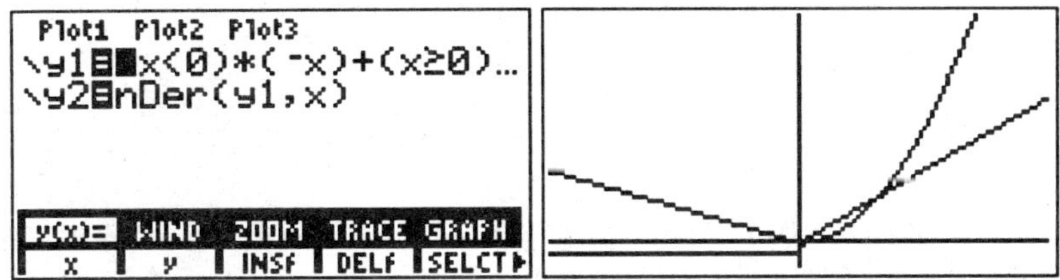

TRACE can be used to find the numerically calculated values of the derivative, but don't expect the correct results near points where the derivative is discontinuous.

The derivative at a given point can be also calculated directly from the graphing menu. Once a graph is plotted, press $\boxed{\textbf{MORE}}$ from the graphing display, and then $\boxed{\textbf{F1}}$ for **MATH**. Then press $\boxed{\textbf{F2}}$ for dy/dx. The cursor appears and can be moved to a point where you want to calculate the derivative. Pressing $\boxed{\textbf{ENTER}}$ gives the derivative at the point of interest. You can also enter the value of x from the keyboard . The result is the derivative value at the cursor location. The example shows the function $f(x) = x^3$.

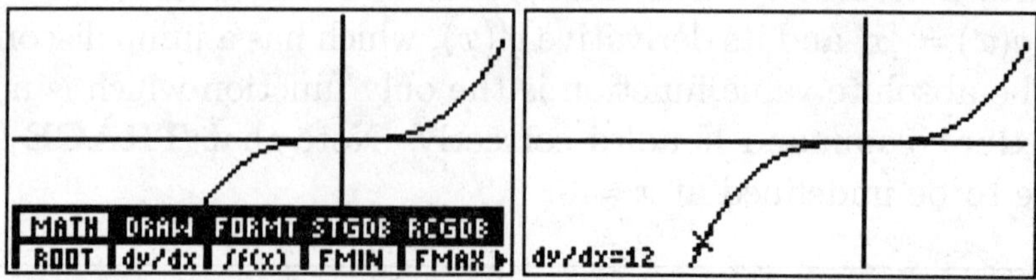

Note that the TI-86 uses the **der1** command in this setting, piecewise defined functions can not be treated in this way. If there is a problem, an **INVALID** error message will occur.

Exercises

1. Plot the function and its derivative in parts a-e. Note the correspondence between the sign of the derivative and the increasing/decreasing nature of $f(x)$. Then plot the function together with its second derivative and note the correspondence between the sign of the second derivative and the concavity of the graph of $f(x)$.

   (a) $f(x) = 1/\left(x^2 + 1\right)$

   (b) $f(x) = \tan^{-1} x$

   (c) $f(x) = x/(x^2 + 1)$

   (d) $f(x) = \sqrt{x} - x$

   (e) $f(x) = (x + 1)^{2/3} x^{3/5}$

2. For the functions in parts a-d, graph the function in the window $[-2, 2] \times [-2, 2]$. Then draw the tangent lines at $x = -1, 0$ and 1. Does the slope of the tangent line increase or decrease? State whether the graph is increasing and concave up, decreasing and concave up, increasing and concave down, or decreasing and concave down.

   (a) $f(x) = e^{-2x/3} - 2 - x$

   (b) $g(x) = e^{2x/3} - 2 + 3x$

   (c) $f(x) = 2 - e^{2x/3} + 2x$

   (d) $g(x) = 2 - e^{-2x/3} + x - 1$

3. Plot the functions $f(x) = x^5 - 2x^3 + 1/10$ and $f(x) = (x^{5/3} - 1/10)(2 - x)^2$ along with their second derivatives in an appropriate sized window. Then use the **ROOT** command to find the inflection points of $f(x)$.

3.6 Linear Approximation of Functions

A graphing calculator can easily be used to illustrate that the tangent line provides a good approximation to a function. As a simple example, consider the function

$$f(x) = e^x$$

It is easy to show that the tangent line to $f(x)$ at $x = 0$ is $y = x + 1$. To examine the linear approximation of $f(x)$ by its tangent line near $x = 0$, use a window with dimensions along the x direction which are close to 0. Two such windows are shown below:

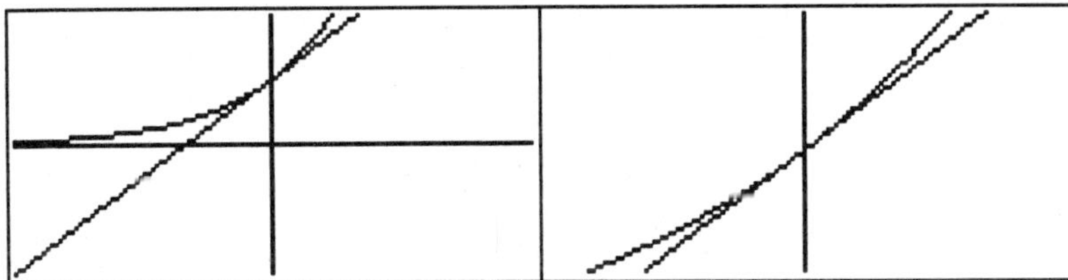

You can see in the second window that the function and its tangent line are essentially coincident near $x = 0$. By repeatedly zooming in you will notice there is always an interval about the point considered that the function and the tangent line look coincident. What you are seeing is the fact that a curve is almost a straight line if we look at a small enough piece.

Exercises

1. Plot a graph of $S(x) = \sqrt{x}$ and its tangent line at $x = 2$. Zoom in near $x = 2$ to get a good view of how the tangent line approximates the square root function near 2. Use the tangent line to approximate the values of $\sqrt{2.01}, \sqrt{2.1}$ and $\sqrt{2.5}$. How do these answers compare with the exact values? Plot a graph of the error made when using a tangent line to approximate the square root near $x - 2$. That is, plot

$$F(x) = |\sqrt{x} - A(x)|$$

where $A(x)$ is the equation of the tangent line at $x = 2$. Over what interval about $x = 2$ is the error less than 0.1?

2. Plot a graph of $S(x) = \sin(x)$ and its tangent line at $x = 0$. Zoom in near $x = 0$ to get a good view of how the tangent line approximates $\sin(x)$ near $x = 0$. Use the tangent line to approximate the values of $\sin(0.1)$, $\sin(0.01)$ and $\sin(0.001)$. How do these answers compare with the exact values? Plot a graph of the error made when using a tangent line to approximate the sine function near $x = 0$. That is, plot

$$E(x) = |\sin(x) - A(x)|$$

where $A(x)$ is the equation of the tangent line at $x = 0$. Over what interval about $x = 0$ is the error less than 0.1?

Chapter 4

Applications of Differentiation

In this chapter we explore some of the many applications of the derivative, with the help of the TI-86 and in conjunction with much of the material in Chapter 4 of Stewart's **Calculus**. In particular, we show how the TI-86 can be used

- to solve max-min problems

- to identifying features of a function's graph, including critical points and inflection points

- to find the zeros of any smooth function using either Newton's method or built-in TI-86 commands

- to analyze motion in the plane by using parametric equations

- to approximate any twice differentiable function.

4.1   Maxima and Minima of Functions

The problem of finding the maximum and minimum values of a given function on an interval can be handled in a number of ways with the TI-86. The simplest approach is to graph the function on the given interval, and locate the maximum and minimum values by observing the graph and using the trace command.

Keep in mind that good results will be contingent on having a good viewing window. Also, remember that the function is only plotted at 126 points, so the location of the maximum and minimum of the function may not correspond to one of the pixel locations. There are a number of ways to remedy this situation. One approach is to use the built-in functions **fMin** and **fMax** which are accessed from the **CALC** menu. These commands approximate the value of x that corresponds to the minimum/maximum value (respectively) of the function on the specified interval. **fMin** and **fMax** are accessed by pressing $\boxed{\textbf{2nd}}$ [CALC], $\boxed{\textbf{MORE}}$ and then $\boxed{\textbf{F1}}$ for **fMin** and $\boxed{\textbf{F2}}$ for **fMax**. The accuracy of the algorithm is set by the value of **TOL** which has a default setting of 10^{-5}. This can be made smaller if you desire by pressing $\boxed{\textbf{2nd}}$ [MEM] $\boxed{\textbf{F4}}$. These functions numerically solve the problem: *find the location of the maximum (or minimum) of $f(x)$ on the interval $[a, b]$.* The syntax for using these commands is

$$\textbf{fMin}(\textbf{f}(\textbf{x}), \textbf{x}, \textbf{a}, \textbf{b})$$

$$\textbf{fMax}(\textbf{f}(\textbf{x}), \textbf{x}, \textbf{a}, \textbf{b})$$

where $f(x)$ is either a formula for the function entered from the keyboard or a reference to a function already entered on the graphing screen, say **y1**. The result of using this command is an approximate value of the x coordinate. If you need the actual maximum or minimum value of the function at this point use the **evalF** command as shown in the screen shot.

In the example shown, the function is $f(x) = 4 - x^2$ and the interval is $[-2, 2]$.

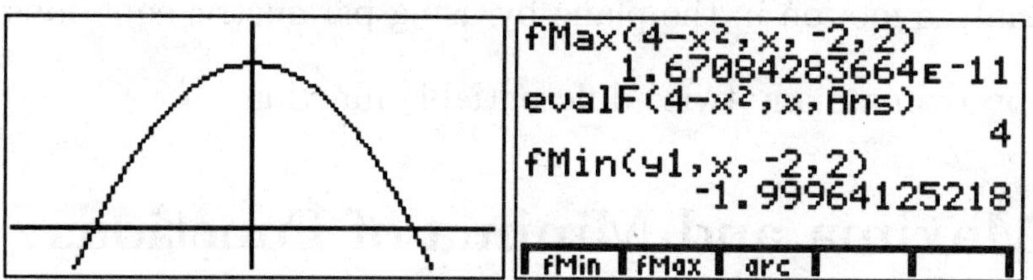

Note that $\boxed{\textbf{2nd}}$ [ANS] retrieves the last answer shown, so you need not enter the long decimal number given. The exact location of the maximum is at $x = 0$ and the minimum occurs at $x = -2$ and $x = 2$. So you see that only approximations are found to these locations, but improvements can be made by decreasing the **TOL** variable. Also be aware that only one answer is returned when **fMin** was used even though there are two values of the x coordinate that give the minimum function value. But note that extrema locations at endpoints or internal points can be found just as easily using this method.

As a final warning, note that even if no maximum or minimum exist on the interval, **fMin** and **fMax** will give an answer. For, example $f(x) = 1/x$ has no maximum or minimum on any interval containing the origin, but an answer is given for each. To address this inadequacy, graph the function you are studying to be sure there is a maximum and minimum.

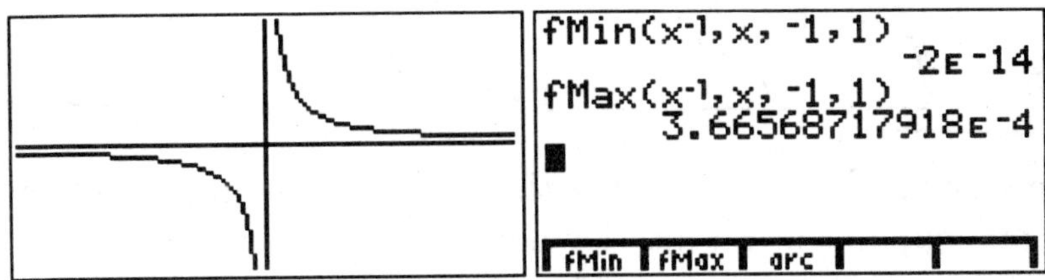

A more graphical approach is also available for finding the maximum and minimum values of a function on an interval. First define the function in the graphing screen, set a viewing window and then graph the function. From the graph screen, press MORE and then F1 (i.e. **MATH**). Pressing F4 or F5 , give **FMIN** or **FMAX** respectively.

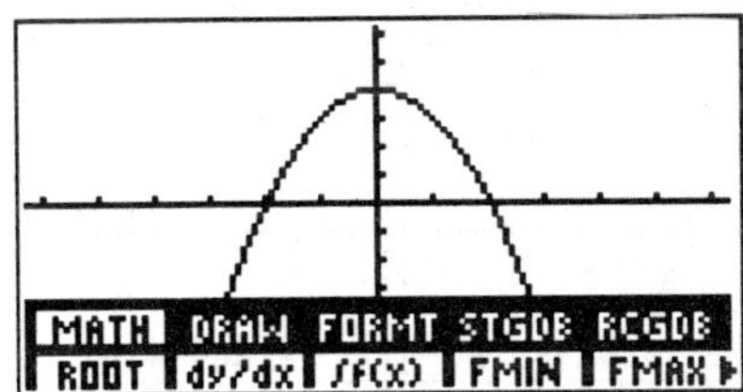

After making your selection, you will be prompted for the left and right endpoints of the interval on which you want to find the maximum or minimum. You can either enter the number from the keyboard or use the arrow keys to move the cursor to the location. Press ENTER after each endpoint is set to its value.

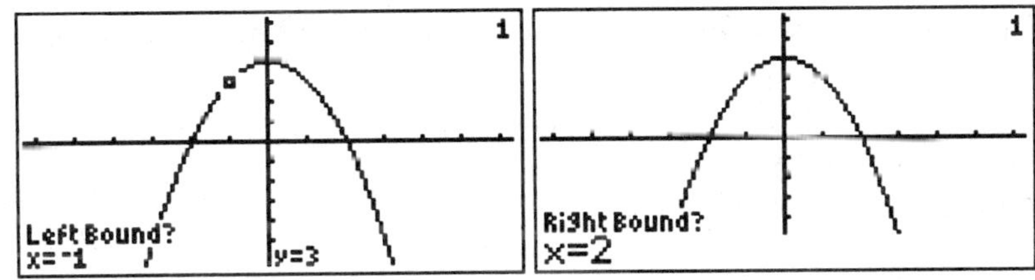

The last entry required is an initial guess which should be a value between the two bounds. Otherwise an error message will result. Press $\boxed{\textbf{ENTER}}$ after entering your guess. The result will be the coordinates of the minimum in this case, with the cursor located on this point.

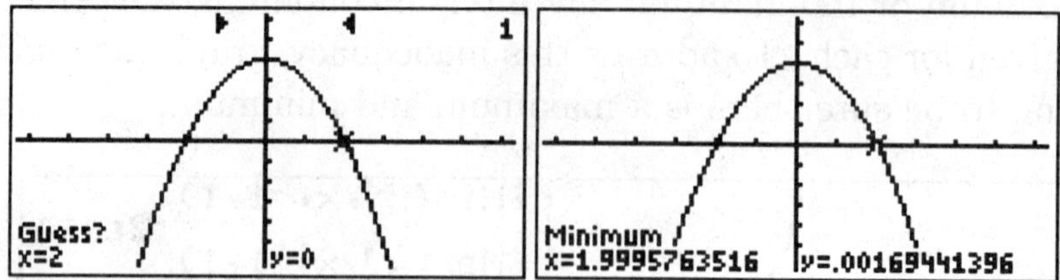

Again care should be taken in using this method for functions that are undefined at an endpoint or an interior point.

For functions that have more than one critical point, use **FMIN** and **FMAX** on each peak or valley by choosing appropriate bounds to set the interval that an answer is desired for. In the screen shot, the function

$$f(x) = 3x + x^2 - 3x^3 - x^4$$

has local maxima on the intervals $(-3, -1)$ and $(0, 1)$.

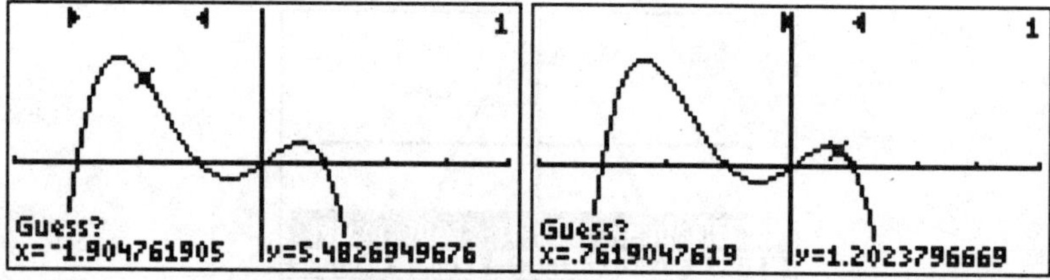

Many times, a max-min problem is posed by a number of sentences that describe the problem. The relevant function extracted from the sentences is the function to be extremized. The physical restraints of a problem often set the interval of study. Once the problem is formulated correctly, a graphical or numerical solution can be found using the methods described above.

As an example, consider the problem of minimizing the surface area of a cylindrical can that has a fixed volume of 10 cm^3. The surface area of such a can is

$$S = 2\pi r^2 + 2\pi rh$$

and the volume is

$$V = \pi r^2 h$$

where r is the radius and h is the height of the can. Solving the volume equation for h and using $V = 10$ in the equation for the surface area gives

$$S(r) = 2\pi r^2 + \frac{20}{r}.$$

The physically relevant part of the domain for $S(r)$ is $r > 0$. After plotting $S(r)$ in a window where the minimum can be seen, **FMIN** can be used to approximate the value of r corresponding to the minimum surface area.

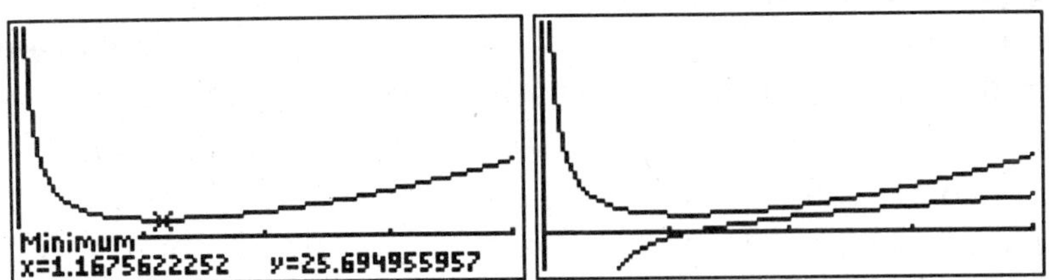

The exact value can be found by solving $S'(r) = 0$ (the second plot above verifies this), which turns out to give $\sqrt[3]{5/\pi}\,cm$. Note that this value can also be found numerically by applying the **ROOT** command to the graph of $S'(r)$.

Exercises: For problems 1-4, find the maximum and minimum value of each function on the given interval.

1. $f(x) = x^3 - x^2 - 42x + 100$ $\qquad$ $[-10, 10]$

2. $g(x) = (x^2 - 5x + 6)e^{-3x}$ $\qquad$ $[-5, 5]$

3. $H(x) = x^3 + \ln(x^2 - 2x + 3)$ $\qquad$ $[-5, 5]$

4. $S(x) = \sin(x \cos x)$ $\qquad$ $[-4, 4]$

5. For problems a-c, use a graphical and numerical approach to solving the given max-min problem.

   (a) A box is to be made from 1000 cm^2 of material. The box will have a square base and an open top. Find the dimensions of the box with the largest volume.

   (b) Find the area of the largest rectangle that can be inscribed inside a circle of radius 10 cm.

(c) An oil pipeline is to be run from an offshore oil well that is located at coordinates (10,5) miles to a city which is located at the origin of the coordinate system. Assume the beach lies along the line $x = 0$. It costs $10000/mile to run the pipeline along the ocean floor and only $7000/mile to run along the beach front. Decide on a pipeline design using only straight lines to connect the oil well and the city that minimizes the cost.

6. An aquarium is to be constructed to hold 20 cubic feet of water. The two ends of the aquarium are to be square, and the aquarium has no top. The glass used for the four sides costs $.50 per square foot, while cheaper glass used for the bottom costs $.35 per square foot. Glue and rubber caulking to fasten and seal the joints between pieces of glass costs $.10 per foot. Finally, framing around the bottom and top perimeters costs $.05 per foot. Find the dimensions of the aquarium that minimize the total material cost.

7. Rework the aquarium problem ignoring all but the cost of the glass.

8. Rework the aquarium problem given that framing around the bottom and top perimeters costs $.50 per foot.

9. Find the area of the largest triangle (with sides parallel to the coordinate axes) that can be contained in the region in the plane bounded by the graphs of $y = 0$, $y = x^2$ and $x = 1$.

10. Find the point on the graph of $y = 1/x^2$, $x > 0$, which is closest to the origin.

4.2   Critical Points and Inflection Points

The critical points of a function give information about the existence of maxima and minima. Since a critical point can be either a point where the derivative vanishes or doesn't exist, the two cases have to be treated differently. From a graphing point of view, enter the function of interest as **y1**, and enter the derivative function as **y2**. If the derivative function is fairly complicated enter either

$$\mathbf{der1(y1, \ x)}$$

or

$$\mathbf{nDer(y1, x)}.$$

Critical points where the derivative vanishes show up as x-intercepts of the derivative graph. Critical points where the derivative doesn't exist, show up as jump discontinuities, or points where there is a vertical asymptote. Recall that **nDer** must be used in cases where the function is piecewise defined.

We now restrict our discussion to the case where the first derivative vanishes. The derivative formula can give exact information about the location of critical points provided one can solve $f'(x) = 0$ analytically. Often this is not the case. For example, the zeros of a fifth degree polynomial cannot be found analytically and we are left to numerical techniques. There are three approaches that can be taken with the TI-86. The first is to use a combination of **TRACE** and **ZOOM** functions to locate the x-intercepts of $f'(x)$. This requires a careful selection of the viewing window. In the screen shot, the derivative function is $f'(x) = x^2 - 3$.

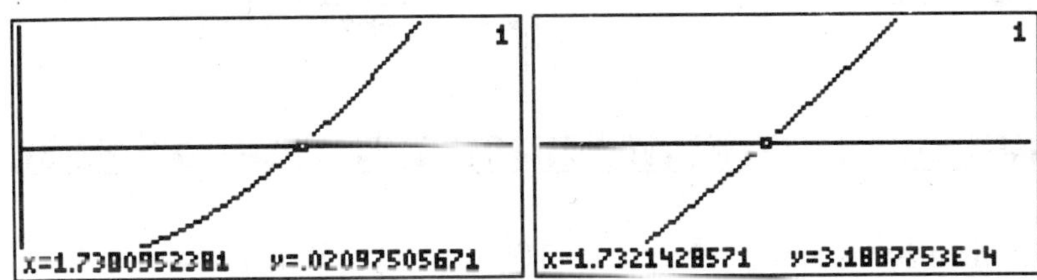

A second approach is to use the **ROOT** command. For this example, take $f(x) = \frac{1}{3}x^3 - \frac{3}{2}x^2$. Once you have graphed the derivative, from the graphing home screen, press $\boxed{\textbf{MORE}}$ and then $\boxed{\textbf{F1}}$ for the **MATH** submenu. Press $\boxed{\textbf{F1}}$ for the **ROOT** command. First use the cursor to select the derivative function. Then either use the keyboard or the arrow keys to enter the interval where you want to find a root.

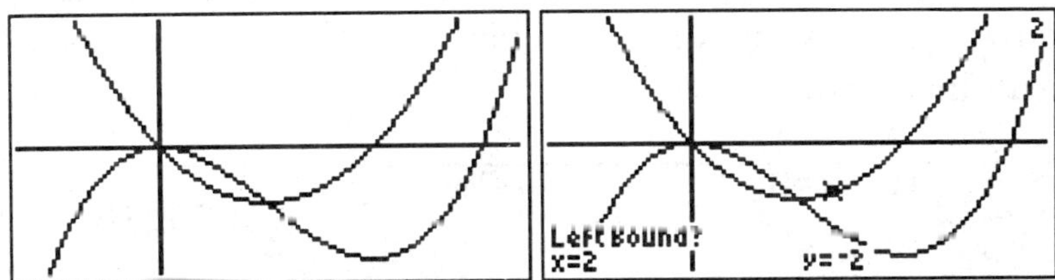

Move the cursor to an initial guess (or enter the value from the keyboard) and then press $\boxed{\textbf{ENTER}}$. The result is the root that is located between the upper and lower settings. The numerical value is shown at the bottom of the display screen.

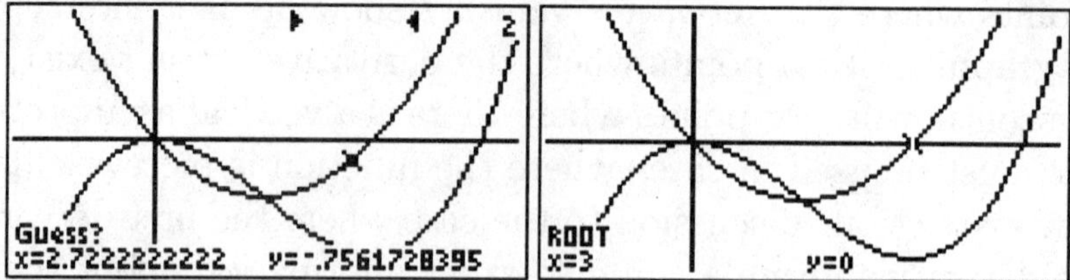

Since the **ROOT** procedure uses the **SOLVER**, an alternative is to use the **SOLVER** directly. Press 2nd [SOLVER]. Then enter the actual derivative formula equals to zero or use the **der1** command. This is shown with the following screen shots.

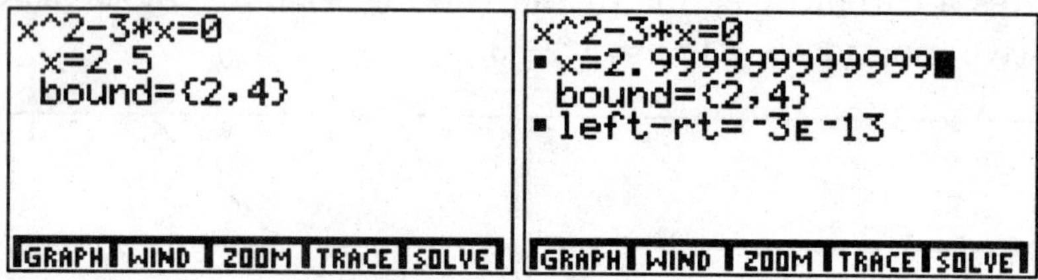

The bound line is initially set to $(-10^{99}, 10^{99})$. You should edit the values in bound to the endpoints of the interval you are interested in. Then set an initial value of x that is inside the defined interval, otherwise you will get a **bad guess** error message. To activate the **SOLVER**, press F5 for **SOLVE**.

The next screen shot shows that the **SOLVER** can be used with either the **der1** or **nDer** derivative commands. Note that **der1** gives the exact result and **nDer** gives an approximate value. This is to be expected as **nDer** uses a numerical approximation to the derivative (although it does have the advantage that it works with piecewise defined functions and other complicated functions).

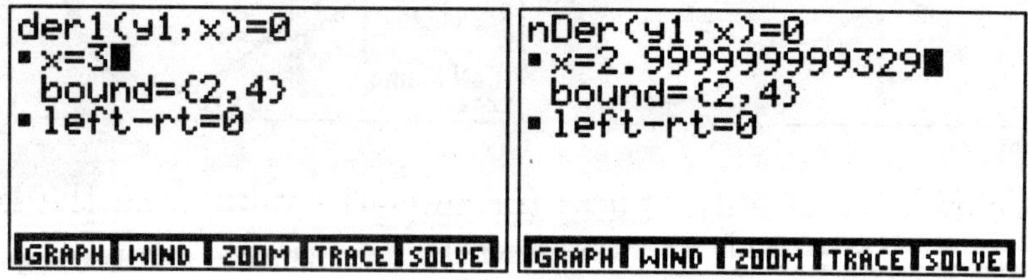

So far, all of the critical points have been points where the first derivative is zero. The other possibility is when the derivative doesn't exist. This case can be handled by plotting the derivative of the function and looking for jump discontinuities or vertical asymptotes. The screen shots below show the two cases for the functions $f(x) = 1 - |x - 2|$ and $g(x) = (x^2 - 1)^{1/3}$. In each case, both the function and the derivative are plotted. The **TRACE** command can be used to help locate these critical points.

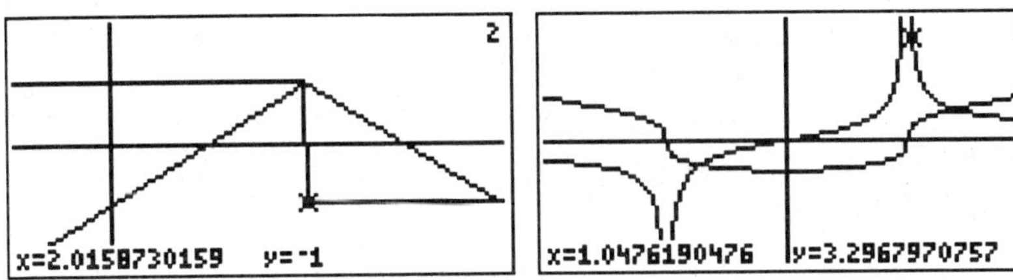

Inflection points are points where the concavity of the function changes from up to down. The second derivative is a measure of the concavity. An examination of the graph of $f''(x)$ tells whether there are inflection points. An inflection point occurs at a point where $f''(x)$ changes sign. In most cases, one looks for roots of $f''(x) = 0$ where $f''(x)$ changes sign. Again it may not be possible to solve $f''(x) = 0$ analytically, so the next best choice is a graphical or numerical approach.

The **TRACE** and **ZOOM** approach described above is one technique for locating these points. Note you may enter the second derivative using either the **der2** command or the exact equation.

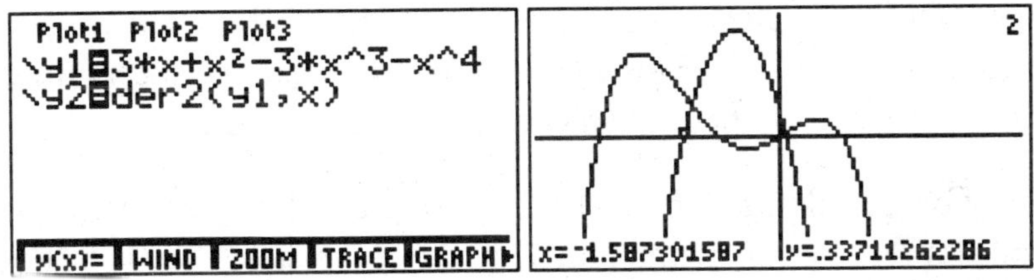

Another approach is to use the built-in function **INFLC** for inflection. From the **MATH** sub-menu of the graphing screen press $\boxed{\text{MORE}}\,\boxed{\text{F3}}$ (for **INFLC**). Again you will prompted for left and right bounds, which can be set with cursor or keyboard. Note that you should be working with the function's graph and not the second derivative as this function finds inflection points of a given function. After giving an initial guess press $\boxed{\text{ENTER}}$. The inflection point coordinates are displayed for you as well as its location on the graph.

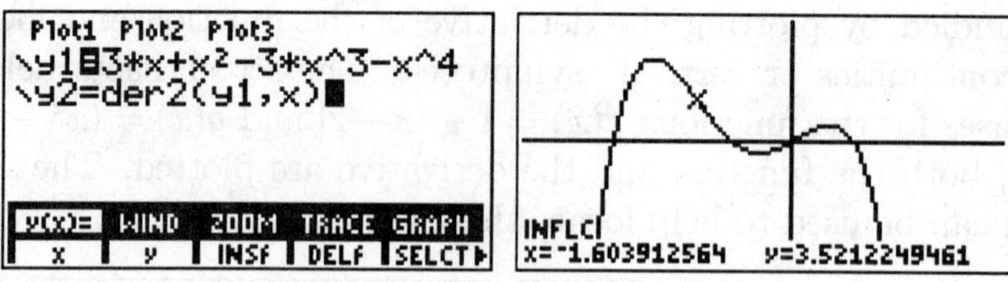

Equivalently, you could plot the second derivative either using the exact formula or the **der2** command. Then the inflection point can be located using the **ROOT** command.

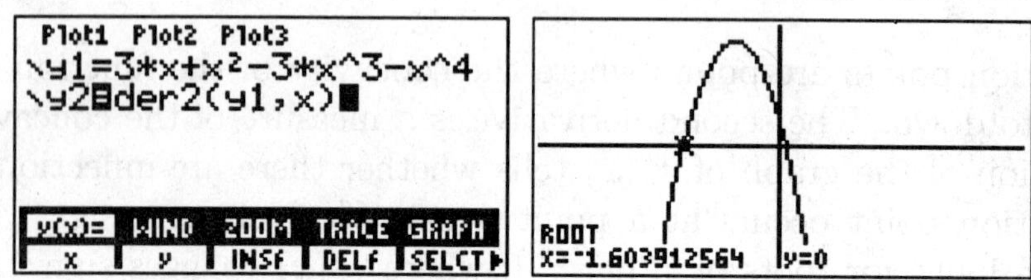

The last way of calculating an inflection point is to solve $f''(x) = 0$ using the **SOLVER**. However, in this case there may be solutions of this equation which are not inflection points. View the graph to be sure of your answers.

As a final note, we should remark that whenever $f(x)$ is a polynomial function, the **POLY** solver can be used to find all of the roots of both $f'(x) = 0$ and $f''(x) = 0$. Then a plot can be used to determine whether these roots are critical points or inflection points of $f(x)$.

Exercises: For problems 1-6 find the location of the critical points and inflection points for the functions on the given interval. Also, list the intervals of increase, decrease, concave up and concave down.

1. $f(x) = x^3 - x^2 - 42x + 100$　　　$[-10, 10]$

2. $g(x) = (x^2 - 5x + 6)e^{-3x}$　　　$[-5, 5]$

3. $H(x) = x^3 + \ln(x^2 - 2x + 3)$　　　$[-5, 5]$

4. $S(x) = \sin(x \cos x)$　　　$[-4, 4]$

5. $f(x) = |x^2 - 1|$　　　$[-2, 2]$

6. $S(x) = |\sin x|$　　　$[-3\pi, 3\pi]$

4.3   Newton's Method

Newton's method is an algorithm which can be used to solve equations of the form $f(x) = 0$. The TI-86 **SOLVER** uses some variation on Newton's method. However, there are situations where you might like to see the details of the calculation.

The basic idea behind Newton's method is to approximate a function by its tangent line at a given point $(x_0, f(x_0)$.

$$y = f'(x_0)(x - x_0) + f(x_0)$$

Then solve the resulting equation for the x intercept by setting $y = 0$. Call this value x_1.

$$0 = f'(x_0)(x_1 - x_0) + f(x_0)$$

The screen shot shows the example of $f(x) = x^2 - 2$ with a starting value $x_0 = 2$. **TRACE** has been used to locate the x-intercept.

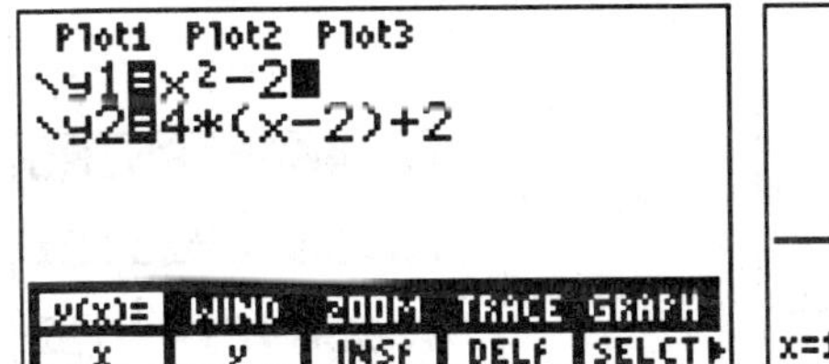
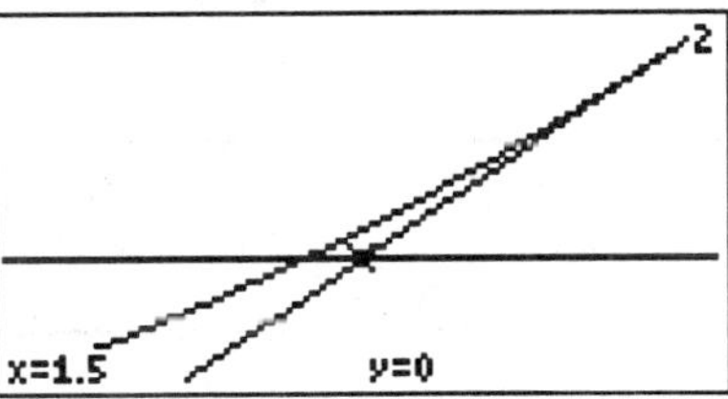

This process can be repeated indefinitely and for many situations the sequence of x intercepts converges to the exact answer. The general Newton's method algorithm is given below:

$$\text{Guess } x_0. \text{ Then compute}$$
$$x_{n+1} = x_n - f(x_n)/f'(x_n) \quad \text{for } n = 0, 1, 2, 3, \ldots$$

The starting value x_0 can be found by using the **TRACE** command on the function's graph. This algorithm will converge provided $f(x)$ is continuously differentiable, $f'(x)$ is nonzero at the true root, and x_0 is sufficiently close to the true root.

Newton's method can be very easily implemented from either the home screen or by writing a short program to automate the process. To implement Newton's method from the home screen, enter the function as **y1** and the derivative as **y2**. If the derivative is a complicated formula, enter **der1(y1,x)**

or **nDer(y1,x)** for **y2**. The follow screen shots demonstrate the case when
$f(x) = x^2 - 2$.

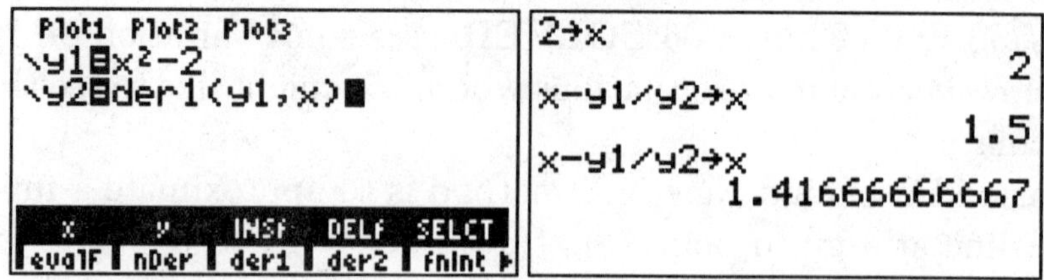

Note that an initial guess is first stored in x, and then successive iterates are
stored in x. The $\boxed{\textbf{2nd}}$ [**ENTRY**] key can be used to save time. Continuing
this procedure leads to excellent approximations to $\sqrt{2}$.

Newton's method does not work with all functions. A typical exception
is when $f(x)$ has a vertical tangent at its root. A simple example function is
$f(x) = (x^2 - 1)^{1/3}$ which has zeros and vertical tangents at ± 1. The screen
shot below verifies that the graph has vertical tangents at ± 1. Iterations
performed with Newton's method oscillate wildly. The first few values are
shown below.

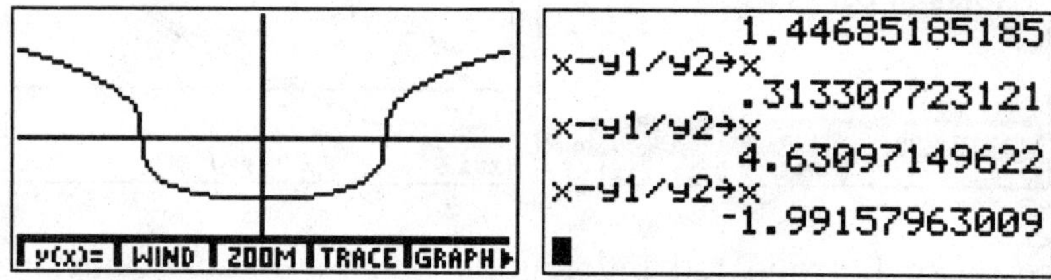

Again **y1** and **y2** refer to the function and its derivative.

Exercises:

1. Use Newton's method to find the cube root of 2, accurate to 6 decimal
   places.

2. Solve the equation $\tan x = x$ using Newton's method on the interval
   $[0, 10]$. Plot an appropriate graph to find out how many solutions you
   need to look for.

3. Write a program to implement Newton's method on your calculator (see
   chapter 9 for help with programming). Your program should input the
   function, a tolerance for stopping the loop, and an initial guess.

4. Compute π to 6 decimal places by finding the zero of the function $f(x) = \tan(x/4) - 1$ between $\pi/2$ and $3\pi/2$. Start with an initial guess of 3.

5. Compute e to 6 decimal places by finding the zero of the function $g(x) = \ln x - 1$.

4.4 Quadratic Function Approximation

We saw in the last chapter how a function can be approximated by a linear function at a point, provided the function is differentiable at that point. We can do much better by using a quadratic function. If a function is twice differentiable at a point then $f(x)$ can be approximated by

$$Q(x) = f(x_0) + f'(x_0)(x - x_0) + \frac{1}{2}f''(x_0)(x - x_0)^2.$$

Note that the first two terms are just the tangent line approximation. The last term, which makes this function quadratic, accounts for the concavity of the function. In fact we refer to the second derivative as being a measure of the concavity of the function's graph. For example, a linear function has zero second derivative and it has zero concavity. Of course if the function is itself a quadratic function, the approximation will be exact.

As an example, consider $f(x) = e^x$. Let's find the equation of the quadratic approximation to $f(x)$ at $x = 0$. Since all derivatives are also e^x, the quadratic approximation is given by

$$Q(x) = 1 + x + \frac{1}{2}x^2.$$

The screen shot shows the function and its approximation on the interval $[-2, 2]$.

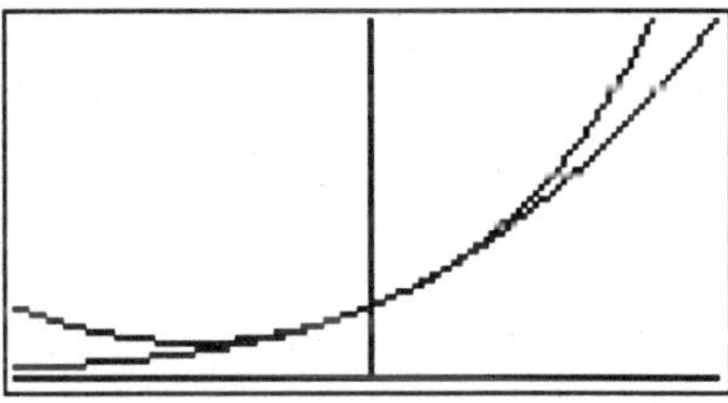

Notice that $Q(x)$ does a very good job of approximating the exponential function near zero. If **ZOOM** is used to examine the behavior near zero, we get a better idea of the accuracy of the quadratic approximation.

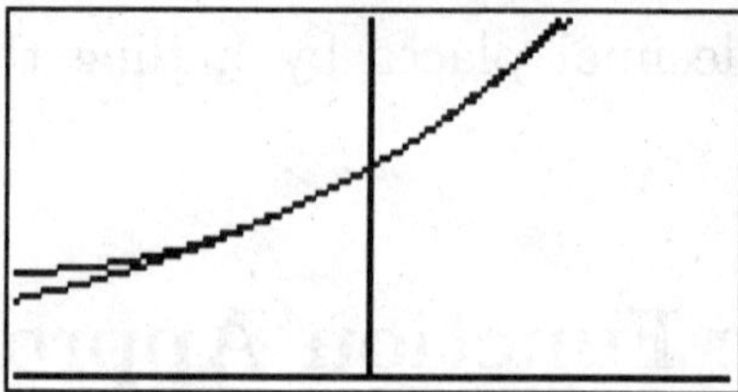

Exercises: For problems 1-4, find linear and quadratic approximations to the given function at the given point. Then plot the function along with its linear and quadratic approximations. Finally, determine the largest interval on which the quadratic approximation lies within 10^{-2} of the given function.

1. $f(x) = \sin x$ at $x = \pi/2$

2. $g(x) = x^3 - 5x^2 + 7x - 9$ at $x = 0$

3. $h(x) = \ln x$ at $x = 1$

4. $k(x) = x^2 + 1/x^2$ at $x = 2$

4.5   Simple Kinematics

The motion of a particle travelling along a line can be represented by a single function of time, $x(t)$, where $x(t)$ gives the position of the particle at time t. For example, $x(t)$ might represent a car travelling down the highway. The velocity of the particle is the time rate of change of the position, and the acceleration is the time rate of change of the velocity.

As an example suppose the position of the particle is given by

$$x(t) = t^3 - 5t^2 + 6t$$

Then the velocity and acceleration are

$$v(t) = \frac{dx}{dt} = 3t^2 - 10t + 6$$

$$a(t) = \frac{dv}{dt} = \frac{d^2x}{dt^2} = 6t - 10$$

These three quantities can be plotted by either using the exact formulas or by using the **der1** command. Unfortunately, you are stuck with using x and y as independent and dependent variables and cannot use the more appropriate t variable for this type of graph.

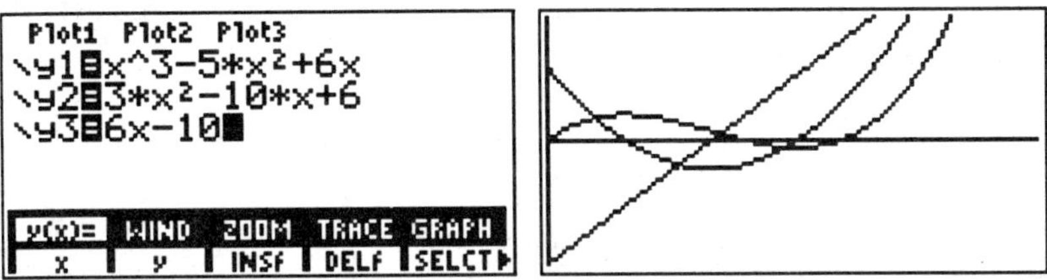

These graphs can be used to find many features of the motion. For example the speed is zero when the velocity graph intersects the horizontal axis, and the particle is reversing its direction of travel when $x'(t) < 0$. In addition, many of the things discussed earlier in this chapter can be used in the study of motion. For example, maximum and minimum acceleration or velocity can be easily found.

The average velocity on the time interval $[a, b]$ is given by

$$v_{AVG} = \frac{x(b) - x(a)}{b - a}$$

You should recognize this value as the slope of the secant line to the position curve. v_{AVG} is the constant velocity a particle can travel along a line to get from the point $(a, x(a))$ to $(b, x(b))$ in the same time that it takes the actual particle along its path. The screen shot shows the given position and the position of a like particle travelling at this constant average velocity between $t = 0$ and 4.

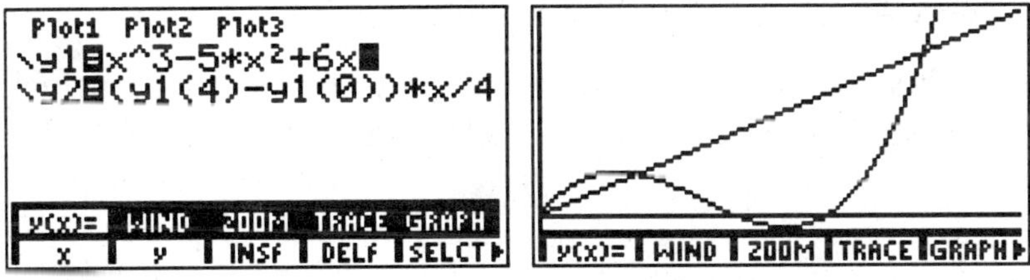

For motion in the plane, two functions of time are required to specify the x and y coordinates of the particle. While these can be plotted the same way as for motion in a line, it makes more sense to use a parametric plot to visualize the motion. When you plot a curve given parametrically, the t

variable is suppressed and a curve is traced out in the plane. This gives a nice simulation of the actual motion. As you see the pixels being turned on the screen, you can imagine the motion of the particle as it travels through the plane.

A simple motion due to a constant acceleration in the vertical direction is given by the equations

$$x(t) = x_0 + v_0 t$$

$$y(t) = y_0 + w_0 t - \frac{1}{2} g t^2$$

where g is the downward constant acceleration. This is exactly the case for the motion of bodies near the earth's surface where g is the gravitational acceleration. The point (x_0, y_0) is the starting position of the particle at $t = 0$ and (v_0, w_0) is the initial velocity in the x and y directions. As an example take $(x_0, y_0) = (0, 1)$ and $(v_0, w_0) = (1, 2)$. Be sure to first set the graphing mode for parametric equations (**Param**) by pressing $\boxed{\text{2nd}}$ [**MODE**]. To simplify, take $g = 10$ m/sec^2.

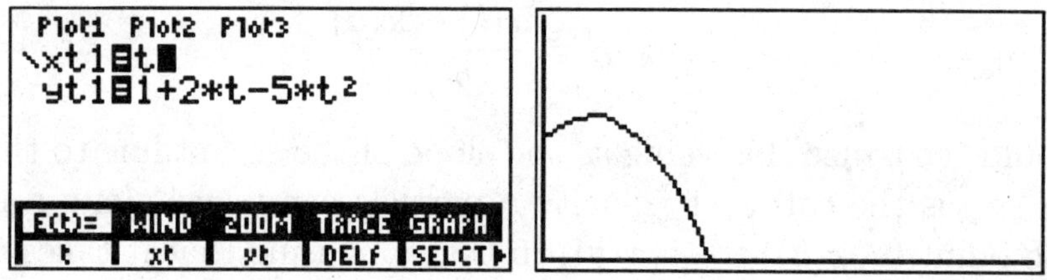

You will notice the familiar shaped parabolic arc that you have seen many times when throwing a ball. The velocity and acceleration can be examined in much the same way as is done for motion along a line.

Exercises: For each position function, plot the position, velocity and acceleration on the indicated interval. Then give the average velocity on the interval.

1. $x(t) = \frac{t-3}{(t-3)^2+1}$ $0 \le t \le 6$

2. $x(t) = e^{-t/2} \sin \pi t$ $0 \le t \le 6$

3. $x(t) = \sin \pi t \cos 5\pi t$ $0 \le t \le 2$

4. For motion in the plane with constant acceleration in the vertical direction as given in the text, plot the position in the plane and the velocity for each set of initial conditions. In each case, find the maximum height the particle reaches and the time at which it occurs.

   (a) $(x_0, y_0) = (0, 1)$ and $(v_0, w_0) = (2, 5)$

   (b) $(x_0, y_0) = (0, 1)$ and $(v_0, w_0) = (1, -1)$

   (c) $(x_0, y_0) = (1, 1)$ and $(v_0, w_0) = (-2, 3)$

5. Show that the curves of motion in the exercise above are parabolas.

Chapter 5

Integration

In this chapter, we examine how areas can be approximated with rectangles, how to define a function in terms of an integral, and how to approximate definite integrals. Related material can be found in Chapter 5 of Stewart's **Calculus**.

5.1   Approximating Areas

While your calculator is able to evaluate definite integrals with a single command, it is instructive to draw rectangles to approximate an area and to examine the limit when the width of the rectangles shrink to zero. In this discussion, we will assume the function considered is either increasing or decreasing on the interval being studied. The reason for this will become evident in a moment. We want to approximate the area between the curve $y = f(x)$ and the lines $x = a$, $x = b$ and $y = 0$. That is, we want to find an approximation to

$$\int_a^b f(x)\, dx$$

using rectangles of equal thickness.

Every calculus text has a picture of these approximating rectangles. If we can define a function that mimics the stairstep appearance of the rectangles, then your calculator can graph it. There is a built-in function on the calculator that has the stairstep appearance. It is usually referred to as the greatest integer function and is defined by the symbol

$$[[x]] = \text{greatest integer} \leq x.$$

For example, $[[4.3]] = 4$ and $[[-0.6]] = -1$. This function can be graphed using the **int** command which is found by pressing $\boxed{\text{MATH}}$, $\boxed{\text{F1}}$ and then $\boxed{\text{F4}}$. The graph of the greatest integer function is shown in **DrawDot** format on a window of size $[-5, 5] \times [-5, 5]$ to show the discontinuities accurately.

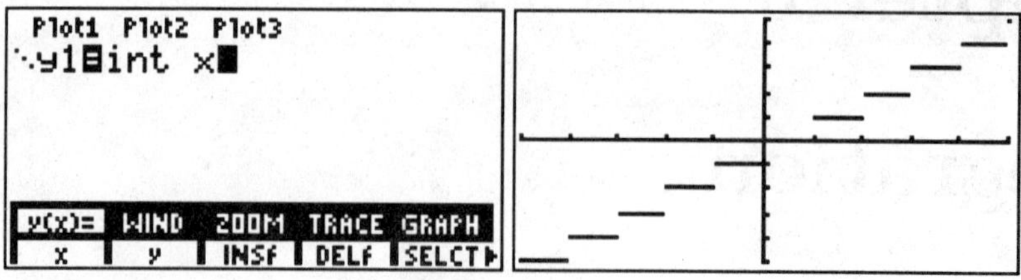

The greatest integer function can be scaled to get any uniform width of stairs with any height desired. By entering $\mathbf{y1} = \frac{1}{4} \, int(4x)$, the width is changed to $\frac{1}{4}$ and the height is changed to $\frac{1}{4}$ (Try it!). In general, taking $\mathbf{y1} = \frac{1}{n} \, int(nx)$ gives n rectangles on each unit interval that increase in step size by $\frac{1}{n}$. Let us define

$$S_n(x) = \frac{1}{n} \, [[nx]]$$

to be this scaled stairstep function. The case when $n = 4$ and 8 are shown.

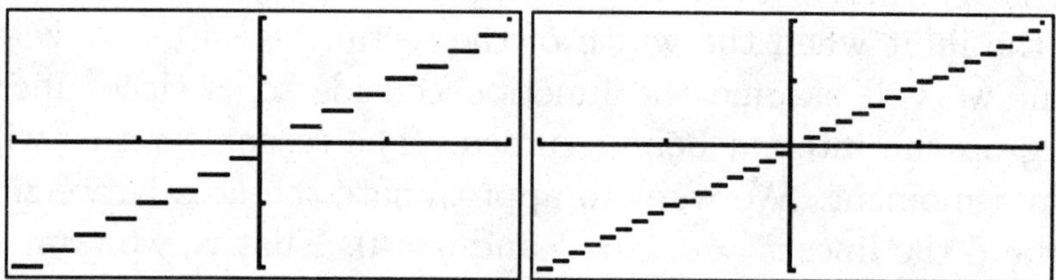

Now we can use this function to plot a graph of the inscribed rectangles under an increasing function. This rectangle function is defined by composing $f(x)$ with $S_n(x)$, that is

$$R_n(x) = f \circ S_n(x) = f(S_n(x)).$$

As simple example, let's approximate the area under $f(x) = x^2$ on the interval $[0, 1]$. To start with, take $n = 8$, so only eight rectangles will be drawn. First, define the functions as shown on the screen shot. Then graph these functions on the window with dimensions $[0, 1] \times [0, 1]$. The result is the desired inscribed rectangles under the function $f(x) = x^2$.

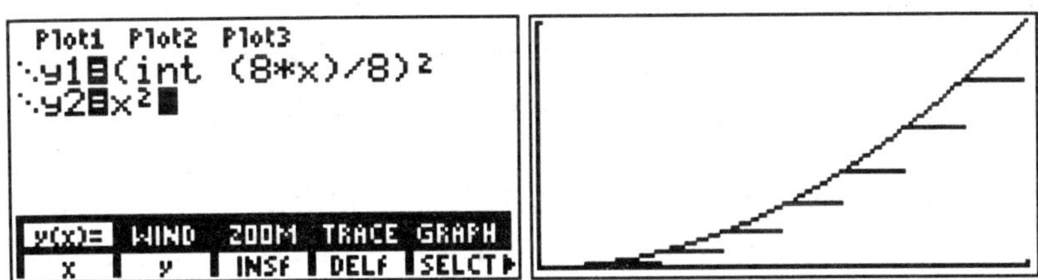

The visual presentation of the approximating rectangles can be made even more pleasing by adding the finishing touch by shading in the rectangles. This is done from the main graphing screen by pressing $\boxed{\textbf{MORE}}$, then $\boxed{\textbf{F2}}$ for **DRAW** and $\boxed{\textbf{F1}}$ for **Shade**. The shade command is of the form

$$\textbf{Shade}(f(x), g(x), a, b).$$

This will shade in the region defined by $f(x) \le y \le g(x), a \le x \le b$. If you enter

$$\textbf{Shade}(0, y1, 0, 1)$$

the result will be the following picture.

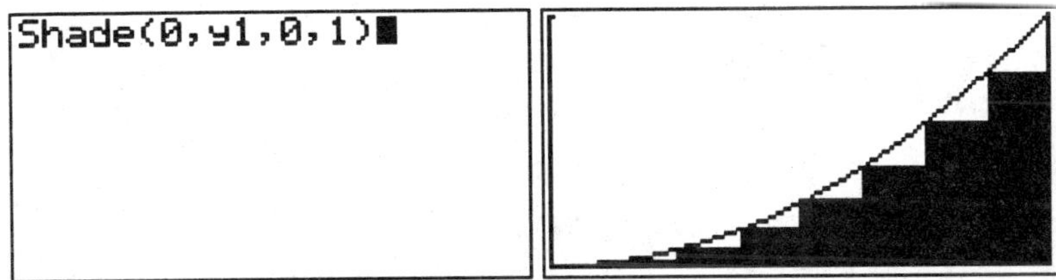

It is easy to look at a refinement of this approximation of the area by editing the definition of the function **y1** and modifying n to a new value. The example is shown in the next screen shot with the 8 replaced by 16, so that 16 rectangles are drawn.

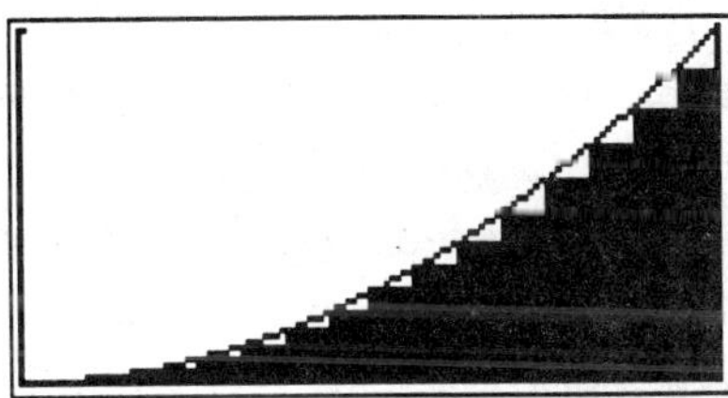

The area under this shaded region can be approximated using the numerical integrator (as will be discussed in the next section) or by using the simple formula

$$\frac{1}{n}\left(f(x_1) + f(x_2) + \cdots + f(x_n)\right)$$

where n is the number of rectangles, x_i represents the point at which the ith rectangle is determined, and $f(x_i)$ gives the height of the ith rectangle. This formula can be computed with $n = 64$ on the TI-86 by using the **sum** and **seq** commands as shown below.

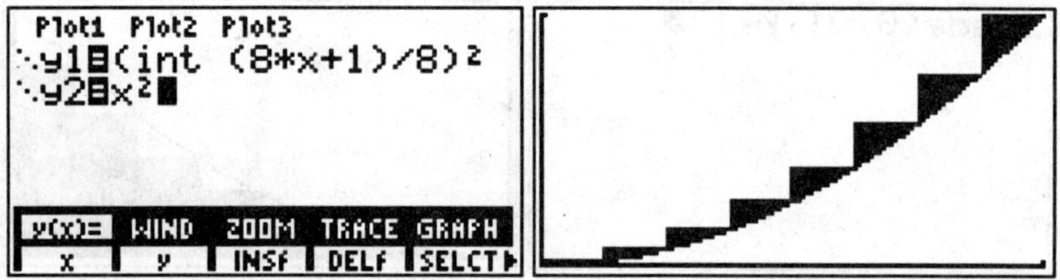

These commands can be entered from the keyboard or by pressing $\boxed{\text{2nd}}$ [CATLG-VARS] $\boxed{\text{F1}}$.

If instead you would like to bound the area above by circumscribed rectangles we replace $S_n(x)$ by $S_n(x + 1)$. Now the composition will read $R^n(x) = f(S_n(x + 1))$. The example shows how to do this for $f(x) = x^2$.

This time the shading has been done between the parabola and the stairstep curve. This is accomplished using the command

$$\textbf{Shade}(y2, y1, 0, 1).$$

Shading the whole rectangles tends to obscure the parabola. This is also an instructive picture since it shows the error made when approximating the parabola's area with these rectangles.

A variation on this procedure can be used to examine decreasing functions. For inscribed rectangles use $S_n(x + 1)$, and for circumscribed rectangles use $S_n(x)$. Inscribed rectangles are shown for the decreasing function

$g(x) = 1/x$ on the interval $[1, 2]$.

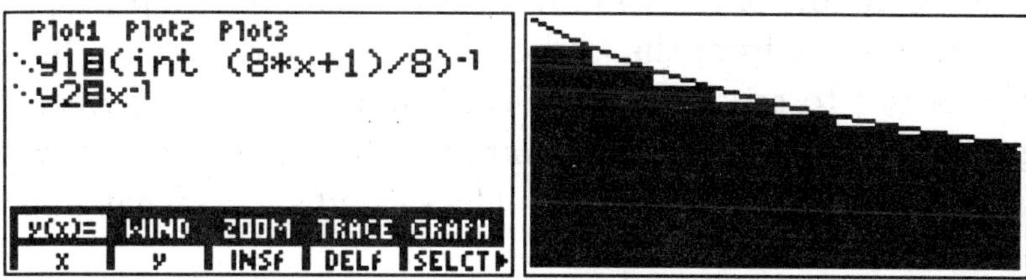

Exercises: For exercises 1-3 use inscribed and circumscribed rectangles to approximate the area bounded by the given function and the x-axis on the given interval. Take $n = 8$, 16, 32 and 64.

1. $f(x) = e^{-x}$ on the interval $[0, 1]$.

2. $g(x) = \sin(x)$ on the interval $[0, \pi/2]$.

3. $f(x) = x^3$ on the interval $[0, 2]$.

5.2 Definite Integrals

The definite integral can be evaluated numerically in two ways with the TI-86 calculator. One approach is done interactively within the **GRAPII** display screen. To evaluate

$$\int_a^b f(x)\, dx$$

first define $f(x)$ as **y1**. Then graph this function on any window dimension that contains the interval $[a, b]$. However, choose wisely if you plan to use the arrow keys and the cursor to locate a and b since they should correspond as close as possible to pixel locations. Select the **MATH** sub-menu and then press $\boxed{\text{F3}}$ for $\int \mathbf{f(x)}$ to evaluate the definite integral.

The example below shows $f(x) = \sin x$ on the interval $[0, \pi]$. The window is set to the dimensions $[0, 6] \times [-2, 2]$.

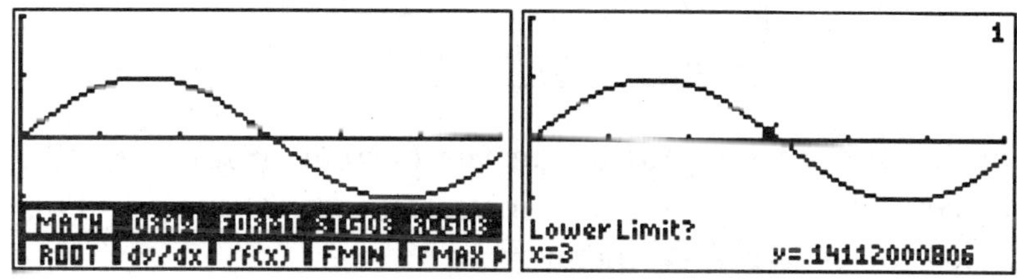

The last screen shows the result of pressing $\boxed{\text{F3}}$. The cursor is set to the value of x in the middle of the window interval and you are prompted to enter the lower limit directly from the keyboard or repeatedly using the arrow keys to move the cursor to the location of the lower limit. The screenshots show the two procedures, with the first showing the cursor as it traces out the graph of $\sin x$ toward the lower limit. The second shows what happens when you type in a lower limit.

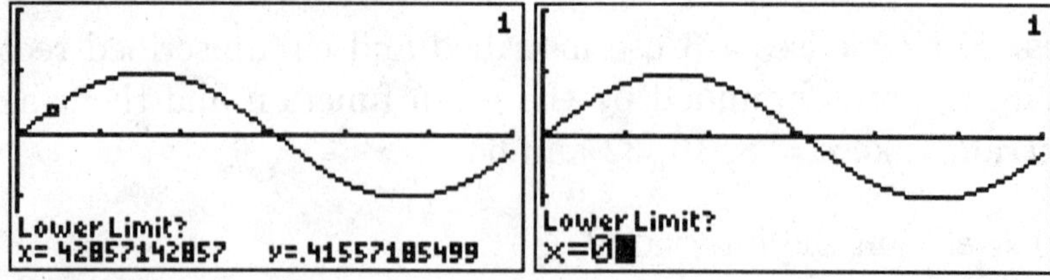

Press $\boxed{\textbf{ENTER}}$ when you have located the lower limit. Then you will be prompted for the upper limit. In this example it is not easy to locate π using the cursor method, since it does not correspond to a pixel on the screen. It is better to enter the exact value to minimize the error in evaluating the integral numerically. There is no need to enter the decimal version of π since you can enter it directly from the keyboard.

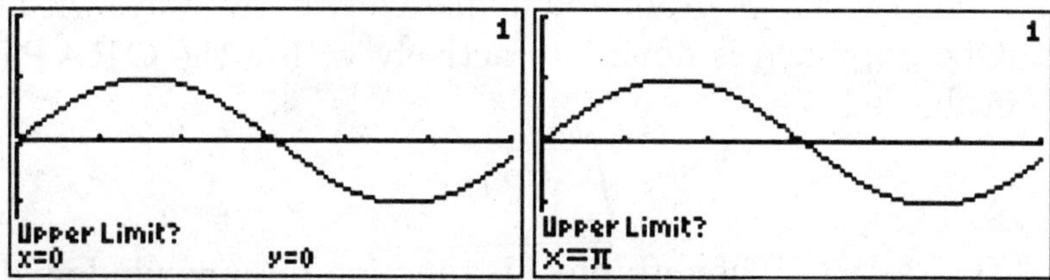

Press $\boxed{\textbf{ENTER}}$ and the numerical value of the integral will be calculated as the area is shaded in on the graph.

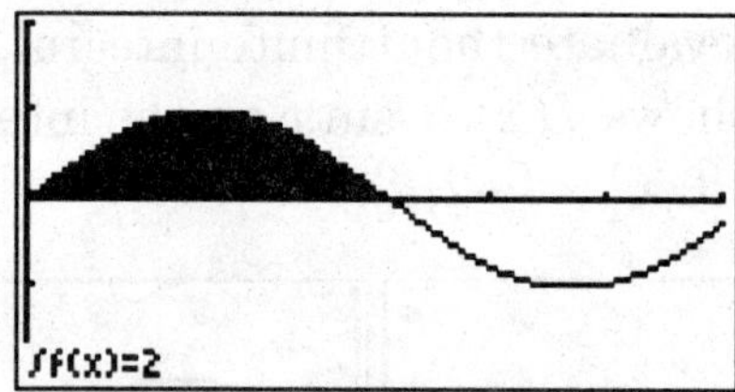

If you plan do another integration on the current graph, use **CLDRW** to clear the screen before entering new lower and upper limits. This is one of the commands in the **DRAW** menu.

This example will show the flexibility in entering limits of integration. Suppose the function is $g(x) = 2 - x^2$ and you want to calculate the area between the x-axis and the function's graph. The limits of integration are easily seen to be $\pm\sqrt{2}$. The limits can be entered directly using the keyboard.

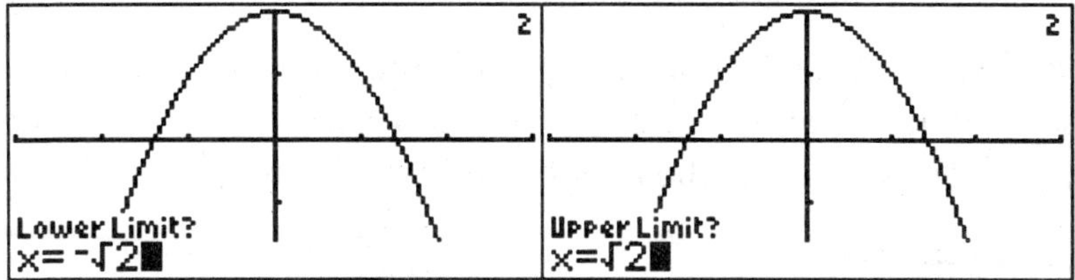

This illustrates a general feature, namely that you can do a calculation using any of the pre-defined functions when entering your limits. This is often helpful in entering accurate values of the integration limits.

The other approach works directly from the standard display screen. Press $\boxed{\textbf{2nd}}$ [CALC] and press $\boxed{\textbf{F5}}$ for **fnInt**, which abbreviates *function integration*. The syntax for this command is

$$\text{fnInt}(\mathbf{f(x), x, a, b})$$

where $f(x)$ is either a function entered directly or one already defined in the **GRAPH** screen. For example,

$$\text{fnInt}(\,\sin \mathbf{x}, \mathbf{x}, \mathbf{0}, \boldsymbol{\pi}) \Leftrightarrow \int_0^{} \pi \sin x \, dx.$$

```
fnInt(sin x,x,0,π)
                   2
fnInt(y1,x,0,π)
                   2
fnIntErr
        1.624311095E-5
```

Note that you can use a function defined in the **GRAPH** screen as the integrand in this command. The accuracy at which the numerical integration can be done is determined by the **tol** parameter. The variable **fnIntErr** is an indicator of the error made. It's value can be found by typing it in using the keyboard or by selecting it from the variables list by pressing $\boxed{\textbf{2nd}}$ [**CATLG-VARS**]. If you suspect the number computed might be wrong, decrease the value of **tol** and see if it makes a difference in the answer.

There are many algorithms for approximating definite integrals, of which two are Simpson's rule and the trapezoidal rule. If you would like to use these techniques, refer to chapter 10 on programming where a program is given for Simpson's rule. A program for the trapezoidal rule can be created by modifying the Simpson's rule program.

Exercises: Evaluate the following integrals numerically, first from the graph screen, and then by using the **fnInt** command from the home screen. Whenever possible, compare with the exact value.

1. $\int_0^{2\pi} \cos^2 t \, dt$

2. $\int_0^1 \frac{1}{(x^2+1)^3} \, dx$

3. $\int_0^{2\pi} \cos^2 t \, dt$

4. $\int_0^1 x \sin^{-1} x \, dx$

5.3   Integral Defined Functions

It is easy to define a function that is defined by an integral such as

$$F(x) = \int_a^x f(t) \, dt$$

or a similar type of integral using the **fnInt** command. Since you can define such a function, you can also graph it, evaluate it, find tangent lines to its graph, or graph its derivatives. Consequently, you should feel as comfortable with these type of functions as the more standard ones.

For example to graph what is known as a Fresnel function,

$$C(x) = \int_0^x \cos\left(\frac{\pi}{2}t^2\right) \, dt$$

go to the graphing screen and enter the equation

$$\mathbf{y1} \; = \; \mathbf{fnInt}(\cos(\pi * t^2/2), t, 0, x)$$

The result can be graphed as shown with window dimensions of $[0, 3] \times [0, 1]$. This function cannot be integrated in terms of elementary functions, so we are stuck with this form. Nevertheless, we can still analyze this function.

In this case, graphing **y1** may take quite a bit of time since each data point is the result of performing a numerical integration. The plot is shown below along the result of using **TRACE**.

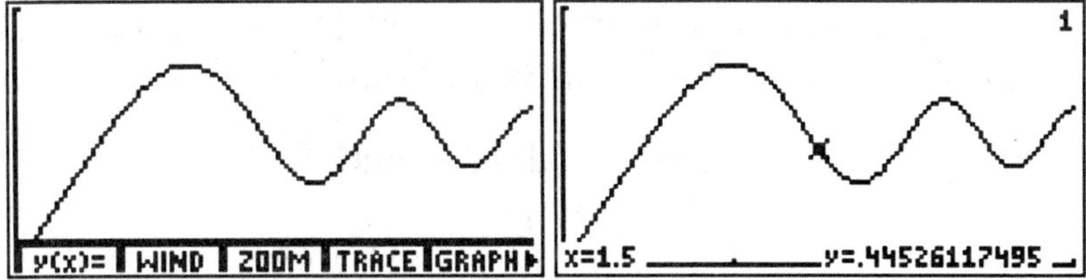

The **TRACE, ZOOM** and **EVAL** commands can be used in the standard way. However, some commands, such as **TANLN** or function integration, do not work correctly for integral defined functions.

It is easy to verify the Fundamental Theorem of Calculus in this case or for any integral defined functions. Enter the following three functions in the **y(x)=** screen:

- the integral defined function

- the derivative

- the integrand

Plotting the latter two will confirm that they are the same function to within small numerical errors. The example shows the excellent agreement between the two functions' graphs. But be warned. Calculating a numerical derivative on a integral defined function can take several minutes to compute and graph.

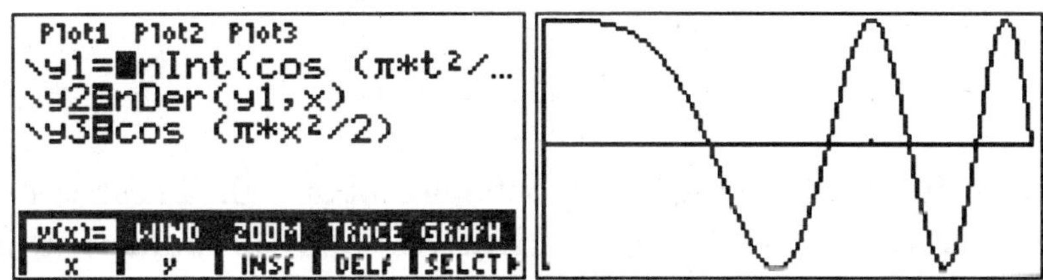

Exercises:

1. Plot the graph of the integral valued function

$$S(x) = \int_0^x \sin\left(\frac{\pi}{2}t^2\right)\, dt$$

2. Plot the graph of the integral valued function

$$L(x) = \int_1^x \frac{1}{t}\, dt$$

The actual value of this function is $\ln(x)$ (see Chapter 7 in Stewart's **Calculus**). Compare the graph you obtained with the graph of $\ln(x)$.

3. Plot the graph of the integral valued function

$$f(x) = \int_{-2\pi}^x \cos\left(t^2\right)\, dt$$

and determine it's maximum and minimum values on the interval $[-\pi, \pi]$.

5.4 Improper Integrals

Improper integrals that have either infinity as an integration limit or an integrand with a mild singularity are not easily handled with numerical integration. For example, you can't enter infinity as one of the limits when you use the **fnInt** command. For integrands with a singularity, this sometimes means a denominator that becomes zero on the integration interval, which could lead to a division by zero error.

First, we discuss improper integrals that have a mild singularity either at an endpoint or at an interior point in the integration interval. If we try to evaluate such an integral using the built-in numerical integrator, we find a mixture of results. For example the integrals

$$\int_0^1 \ln(x)\, dx \;=\; -1$$

$$\int_0^1 \frac{dx}{\sqrt{x}} \;=\; 2$$

are improper due to the integrand's behavior at $x = 0$. These two examples are easily done exactly using anti-derivatives, and the numerical integrator gives reasonably good results.

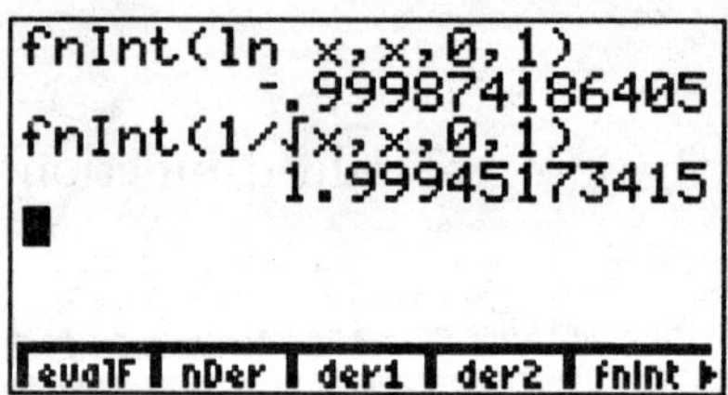

But if we try to evaluate

$$\int_0^1 \frac{dx}{(1-x)^{4/5}} = 5$$

we get a tolerance not met error. In this integral the singularity is stronger than the square root, and the numerical integrator has difficulty meeting error goals.

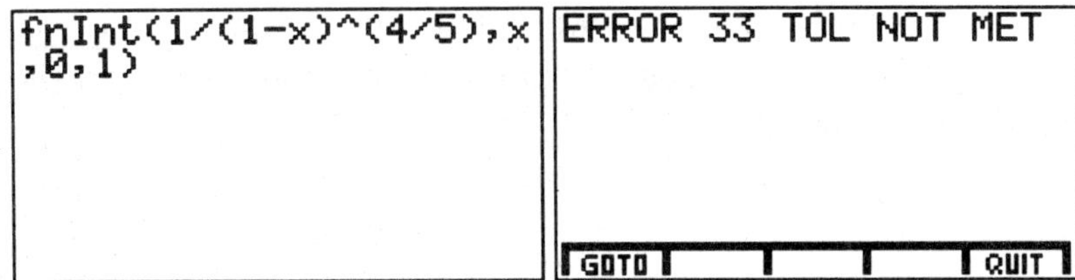

There is no hope for the numerical integrator if the discontinuity occurs inside the integration interval, as a division by zero error message is guaranteed. For example try to evaluate

$$\int_1^1 \frac{dx}{x^{1/3}} = 0$$

where the integrand has a zero denominator at $x = 0$. The result is a division by zero error.

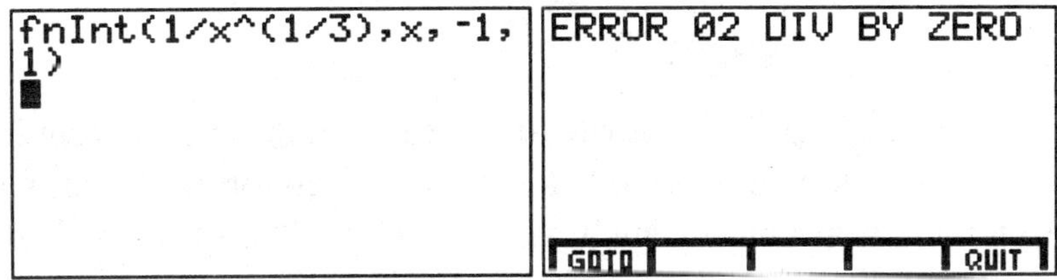

An integral with a singularity at an interior is best divided into two integrals with the singularity at one of the endpoints. For this example, we can write

$$\int_{-1}^1 \frac{dx}{x^{1/3}} = \int_{-1}^0 \frac{dx}{x^{1/3}} + \int_0^1 \frac{dx}{x^{1/3}}$$

Then a good numerical approximation to the answer can be found using

fnInt.

```
fnInt(1/x^(1/3),x,-1,
0)+fnInt(1/x^(1/3),x,
0,1)
                    1E-13
```
```
evalF  nDer  der1  der2  fnInt ▶
```

Another approach to using a calculator to help evaluate improper integrals is to replace the troublesome endpoint where the singularity occurs with a variable. If the problem point occurs inside the interval, divide the integral into two parts, each of which has a problem endpoint. For example, consider the integral defined function

$$F(x) = \int_0^x \frac{dx}{\sqrt{1-x}}$$

We can use numerical integration to evaluate this function for values of x that approach $x = 1$. Successive evaluation of the improper integral in this manner is like finding the limiting value of $F(x)$ as x tends to 1. Once you enter the command **fnInt** use $\boxed{\textbf{2nd}}$ $\boxed{\textbf{[ENTRY]}}$ to repeat the calculation. You may not get the exact value but you have a good approximate value. The screenshot shows a series of calculations that illustrate the limiting behavior to 2.

```
fnInt(1/√(1-x),x,0,.9
999)
            1.98000062737
fnInt(1/√(1-x),x,0,.9
9999)
            1.99367380846
fnInt(1/√(1-x),x,0,.9
9999)
```

Of course, this is not a proof that the answer is 2, but it is a strong indication. A graphical approach would be to take **y1** to be the integral defined function $F(x)$ and plot it on a window such as $[.99, .999]$. The graph will flatten out and approach the line $y = 2$ asymptotically if in fact the improper integral converges.

Not all improper integrals converge and you may not know initially if the integral is convergent. However, if you follow the procedure above, you may see that the numerical calculations do not get close in the limiting sense to a particular number. For example,

$$\int_0^1 \frac{dx}{x}$$

is a divergent integral. Several numerical integrations of the integrand on the interval $[\varepsilon, 1]$ where ε is a small number shows that the values are increasing and do not appear to approach a limit.

```
fnInt(1/x,x,.001,1)
           6.90775533865
fnInt(1/x,x,.0001,1)
           9.21034037548
fnInt(1/x,x,.00001,1)

          11.5129254767
```

The technique just discussed can be used to evaluate improper integrals with one or more endpoints being infinity. For example, to estimate the value of

$$\int_0^\infty t^3 e^{-t} dt = 6$$

numerically, we evaluate the integral defined function

$$F(x) = \int_0^x t^3 e^{-t} dt$$

at values of x that are large. The integration interval can be very large in this case. This may mean that the resulting error is significant. That's why other techniques have been developed for evaluating improper integrals of this type. Using the **fnInt** command gives the following results.

```
fnInt(x^3*e^(-x),x,0,
10)
           5.93798369587
fnInt(x^3*e^(-x),x,0,
20)
           5.99998077759

evalF  nDer  der1  der2  fnInt
```

A graphical approach is also possible. For example, we can plot $F(x)$ on a window with $10 < x < 20$.

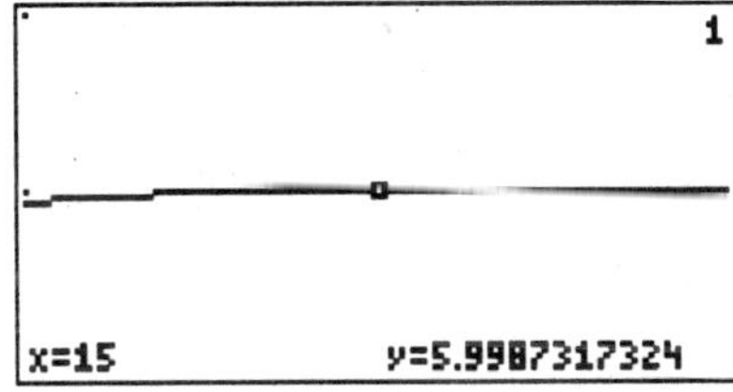

You will note that the graph of this function flattens out and approaches the asymptotic value of 6. In this example we are lucky that the convergence of the integral is fast. In fact, when $x = 20$ the result is within 10^{-5} of the exact value (check this). Of course, this is not always the case. Each improper integral has to be analyzed on an individual basis.

Sometimes a change of variable can be helpful in putting the improper integral into a form that is more easily evaluated numerically. For example, let $t = v^2$ in the last example. Then the integral becomes

$$\int_0^\infty t^3 e^{-t}\,dt = 2\int_0^\infty v^7 e^{-v^2}\,dv$$

which has the advantage that the factor e^{-v^2} decays faster than e^{-t} does. The result of this change of variable is dramatic as shown in the screen shot where the upper limit can be taken to be only 4!

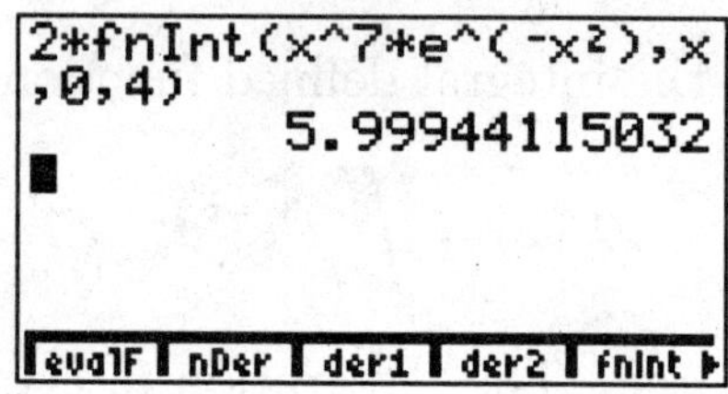

Exercises: Approximate the improper integrals or determine divergence in exercises 1-3.

1. $\int_{-\infty}^{\infty} e^{-x^2}\,dx$

2. $\int_0^\infty \frac{1}{1+x^4}\,dx$

3. $\int_0^1 \frac{1}{x^3+3x^2}\,dx$

4. A function f defined on $(-\infty, \infty)$ is a probability density function if $f(x) \geq 0$ for all x in $(-\infty, \infty)$, and $\int_{-\infty}^{\infty} f(x)\,dx = 1$. Find the number k (at least approximately) so that each of the following is a probability density function. Then graph f.

   (a) $f(x) = \frac{k}{1+x^2}$

   (b) $f(x) = \frac{k}{(1+x^2/2)^{3/2}}$

   (c) $f(x) = ke^{-x^2/2}$

Chapter 6

Applications of Integration

In this chapter we learn how to approximate the area between two curves, how to compute the arc length of a curve, and how to handle volumes and center of mass calculations using the TI-86. The main tools for area and volume calculations are the plotting capabilities, the **SOLVER** and the numerical integration routine **fnInt**. Arc length can be computed using **arc**. Related material can be found in Chapter 6 of Stewart's **Calculus**.

6.1   Area Between Curves

The TI-86 can be used to set up an integration for finding the area between two curves. First, we need to plot the graphs of the two functions $f(x)$ and $g(x)$ to determine whether $f(x) > g(x)$ or $g(x) > f(x)$. This information is important in setting up the integrand correctly. Next we need the intersections of the two functions to determine the limits of integration. The intersections satisfy the equation

$$f(x) = g(x)$$

and can be solved in a number of ways. If the functions are simple enough the solutions can be found exactly using analytical methods. If this is not the case, the intersections can be found by using the **SOLVER**. Once the functions are defined in the **y(x)=** screen, the solver equation is simply

$$y1 = y2$$

If the two functions are graphed first, then **TRACE** can be used to estimate the intersections. These estimates can be used as initial guesses with the **SOLVER** to get more accurate values.

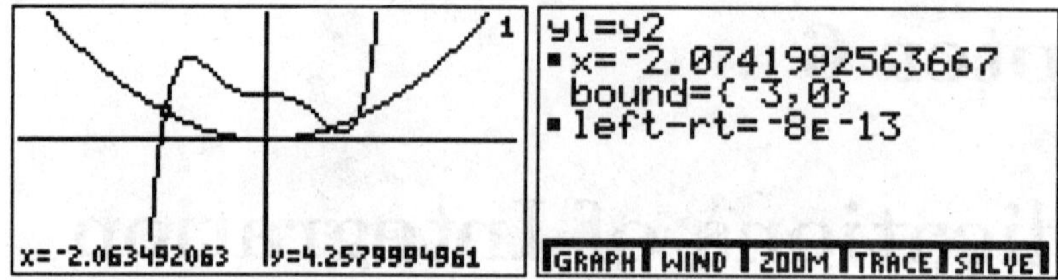

A more graphical approach is to use the built-in routine **ISECT** for finding intersections of two functions. **ISECT** uses the **SOLVER** to do the calculation. For the sake of an example, take

$$f(x) = x^2$$
$$g(x) = x^5 - 4x^3 + 7$$

and define these in the **y(x)=** screen as **y1** and **y2** respectively. Then set the window dimensions to be $[-5, 5] \times [-20, 20]$. After graphing the functions, press $\boxed{\textbf{MORE}}$, $\boxed{\textbf{F1}}$ for **MATH**, and then $\boxed{\textbf{MORE}}$ again. Then press $\boxed{\textbf{F3}}$ for **ISECT**. The **ISECT** routine will first ask which is the first curve and which is the second curve. If there are only two curves drawn this will be easy. If more than two curves are plotted, use the up and down arrows to select the curves. After selecting the first curve press $\boxed{\textbf{ENTER}}$, and again press $\boxed{\textbf{ENTER}}$ after selecting the second curve.

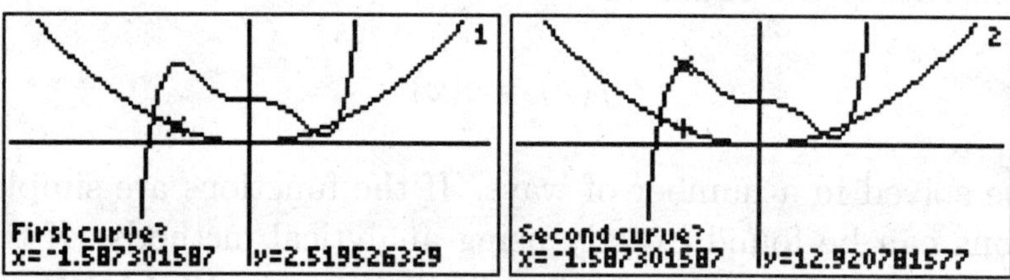

ISECT will then prompt for an initial guess of the intersection of the two curves. Enter an initial guess either from the keyboard or using the cursor keys to locate a point near the intersection. Then press $\boxed{\textbf{ENTER}}$. The result is the coordinates of the intersection point.

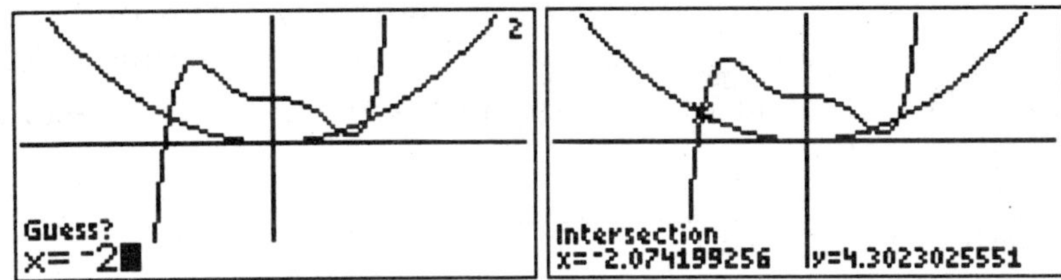

The x-coordinate of the intersection is stored in the **ANS** variable which you will need later in doing the integration. It is a good idea to leave the graphing screen after you find each intersection and move the answer from **ANS** to another memory location.

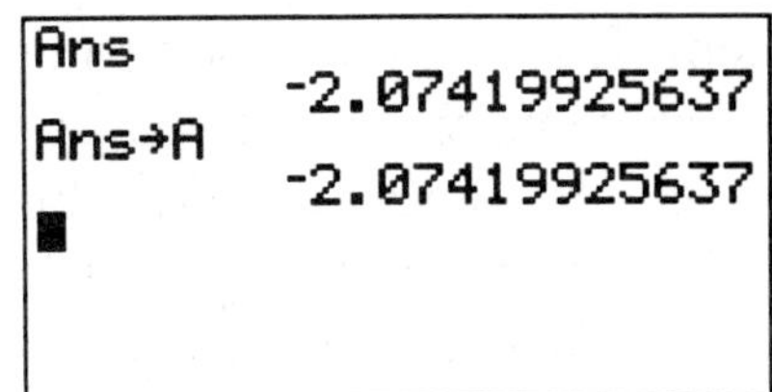

If you plan to use numerical integration then having the intersection values saved in variables makes it easy to use **fnInt**. The remaining two intersections are approximately 1.328 and 1.856.

A final approach to finding the intersections is to use a combination of **TRACE** and **ZOOM** to find the intersections to the desired accuracy. Using the **ZIN** command zoom in about the intersection points successively getting better approximations to the intersection point. Use the **TRACE** to isolate the intersection.

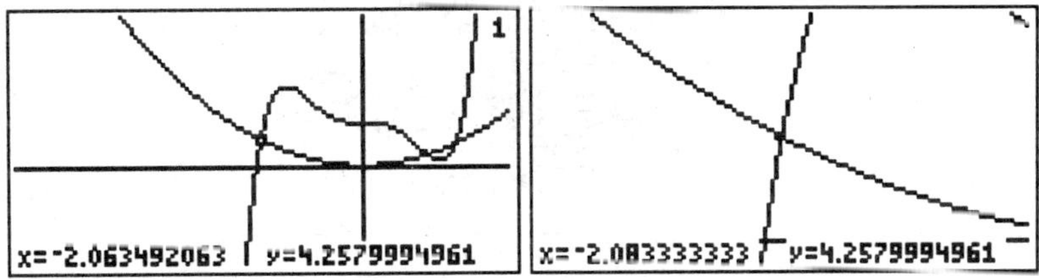

This approach is the only one when working with parametric curves as there is no **ISECT** command in this graphing mode.

If you like you can shade in the area that is to be found using the **Shade** command. The syntax for the command is

$$\textbf{Shade}(f, g, a, b)$$

where f and g are functions either given explicitly or defined in the graphing screen, and a and b are the left and right endpoints. The function should satisfy the inequality $f(x) < g(x)$ on $[a, b]$. The **Shade** command is accessed from the **DRAW** submenu of the graphing screen. For this example the desired area is shaded in.

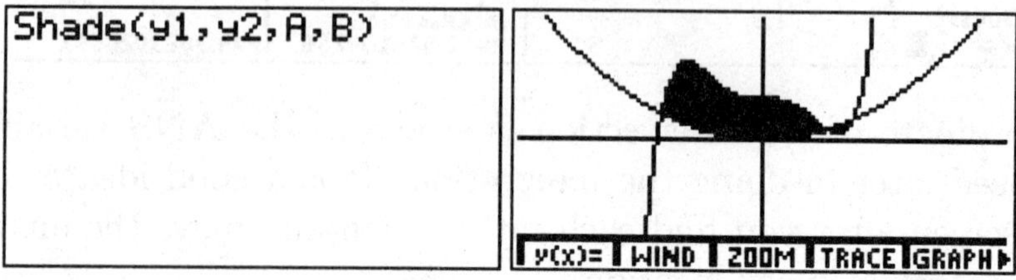

To complete the problem, **fnInt** can be used to numerically approximate the desired area. In this problem, an anti-derivative exists for $f(x) - g(x)$ so a numerical result can also be found using the Fundamental Theorem of Calculus. But since the integration limits are quite complicated, numerical integration is preferable. Since there are three intersections, the area between them is the sum of two integrals

$$A = \int_a^b (g(x) - f(x))\, dx + \int_b^c (f(x) - g(x))\, dx$$

as shown in the screen shot. This integral is computed below, followed by an alternative approach of integrating $|f(x) - g(x)|$ (so that we don't have to worry about the reversal in the sign of $f(x) - g(x)$).

```
fnInt(y2-y1,x,-2.074,
1.328)+fnInt(y1-y2,x,
1.328,1.856)
           23.6133829558
fnInt(abs (y1-y2),x,-
2.074,1.856)
           23.6135072211
```

Exercises:

1. Find the area between the curves $y = e^{-x^2}$ and $y = x^2$.

2. Find the area of the region that is bounded above by the curves $y = 7\ln x$ and $y = 4 - x - x^3$ and below by the x-axis.

3. Find the area between the two curves $y = \sin x$ and $y = \frac{1}{2}\sin 2x$ on the interval $[0, 2\pi]$.

4. Find the area of the region that is bounded by the y axis, and the curves $y = e^{-x}$ and $y = 2x$.

5. Find the area between the curves $y = \sqrt{2 - x^2}$ and $y = 1/(1 + x^2)$.

6.2 Arc Length

The length of the curve $y = f(x)$ for $a \le x \le b$ is given by the integral

$$\int_a^b \sqrt{1 + f'(x)^2}\,dx$$

In most cases, this leads to an integrand that can be fairly complicated. In fact, very few arc length calculations can be done using pencil and paper, so a numerical approach is desired.

To evaluate this integral numerically on the TI-86, we use the **arc** command which is accessed from the **CALC** menu. After pressing $\boxed{\textbf{2nd}}$ [CALC], press $\boxed{\textbf{MORE}}$ and then $\boxed{\textbf{F3}}$ for **arc**. The syntax of this command is

$$\textbf{arc}(f(x), x, a, b)$$

where the first argument is an actual function or a **y1** type definition, the second argument is the variable name, and remaining two arguments are the integration limits.

For example, consider $f(x) = x^2$ on the interval $[0, 2]$. If the function has been defined in the **y(x)=** screen as **y1**, then the function's name can be substituted for its formula.

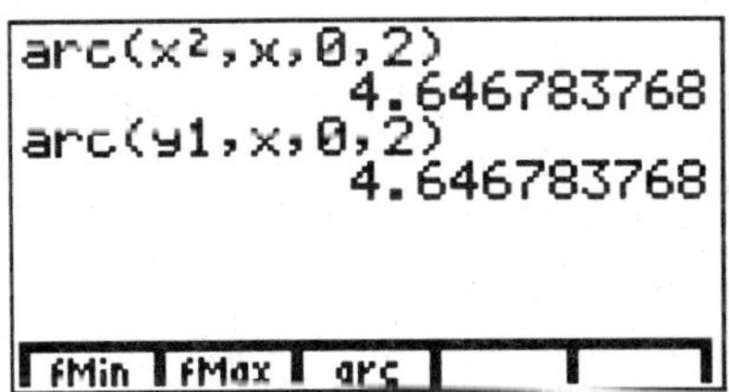

The calculation can also be done interactively from the graphing display from the **MATH** menu. After selecting the **MATH** menu, the **ARC** command is run by pressing $\boxed{\textbf{MORE}}$ once and then $\boxed{\textbf{F5}}$. Use the arrow keys or enter a number from the keyboard to set the left endpoint and then press $\boxed{\textbf{ENTER}}$, and then again use the arrow keys or keyboard to set the right-hand endpoint. After pressing $\boxed{\textbf{ENTER}}$ the arc length is calculated as shown.

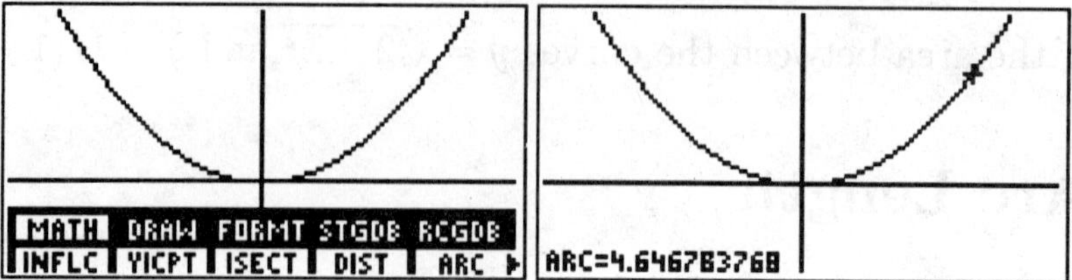

A similar procedure holds for parametric equations given by $x = f(t)$, $y = g(t)$ with $a \leq t \leq b$. In this case, the arc length is given by

$$\int_a^b \sqrt{f'(t)^2 + g'(t)^2}\,dt.$$

There are two ways to evaluate this integral with your calculator. One approach is to use the **fnInt** command with the integrand as shown, as there is no equivalent command like **arc** for parametric equations. Note that this integrand can be quite complicated so it might be easier to use the **der1** command to form the integrand.

As an example take $f(t) = e^t$ and $g(t) = t$ on the interval $[0, 1]$. The derivatives are easily calculated, but **der1** can also be used to calculate the derivatives. This is shown in the screen shot below.

```
fnInt(√(der1(e^t,t)²+
der1(t,t)²),t,0,1)
          2.00349711163
```

As a final example, let's find the perimeter of a circle using the built-in integrator. After setting the graphing mode to **Param** in the **MODE** screen, enter the following functions for **xt1** and **yt1**:

$$\begin{aligned} x(t) &= \cos t \\ y(t) &= \sin t \end{aligned}$$

This corresponds to a parameterization of a circle with a radius of 1 centered at the origin. The window should be set to $tMin = 0$ and $tMax = 2\pi$. To evaluate any arc length from the parametric **GRAPH** screen, press $\boxed{\text{MORE}}$, $\boxed{\text{F1}}$ for **MATH**, and then $\boxed{\text{F5}}$ for **ARC**. You will be prompted to enter the value of the endpoints directly from the keyboard or by using the arrow keys to locate the point. The result shows the familiar perimeter.

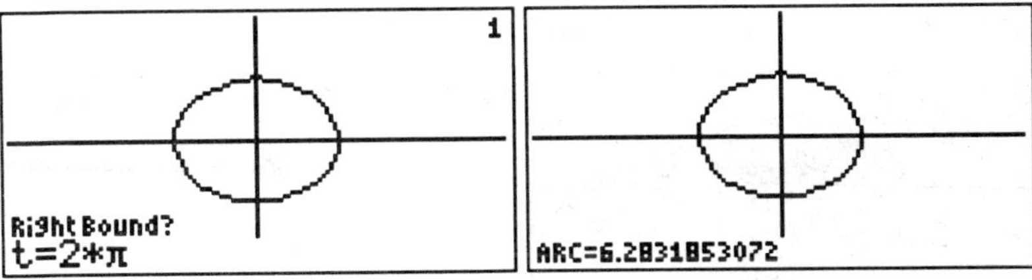

Exercises:

1. Find the arc length of the function $f(x) = x^3$ on the interval $[-1, 1]$.

2. Plot the cycloid defined parametrically by $x(t) = 2t - 2\sin t$ and $y(t) = 2 - 2\cos t$. Find the length of the cycloid for $0 \le t \le 2\pi$.

3. Find the length of the sine curve $y = \sin x$ over one period.

4. Find the arc length of the hypocycloid

$$x(t) = 4\cos^3 t, \, y(t) = 4\sin^3 t$$

for $0 \le t \le 2\pi$.

5. Find the arc length of the $\arctan x$ curve for $0 \le x \le 1$.

6.3 Volumes

One of the many applications of integration is finding volumes of three dimensional objects. While it may not be possible to plot three dimensional objects with a graphing calculator without writing an elaborate program, it is possible to graph cross-sections and evaluate the associated integrals numerically. The basic volume integral uses the cross-sectional area $A(x)$ as the integrand. This gives

$$V = \int_a^b A(x)\, dx$$

Consider the example of a curved triangular base defined by the equations $y = \pm(1 - x^2)$ and equilateral triangle cross-sections. The area of each cross-section is half the base times the height, and will be

$$A(x) = \frac{1}{2}2(1 - x^2)\sqrt{3}(1 - x^2) = \sqrt{3}(1 - x^2)^2.$$

The integrand is simple enough that it can done exactly using anti-derivatives or numerically using the **fnInt** command as shown.

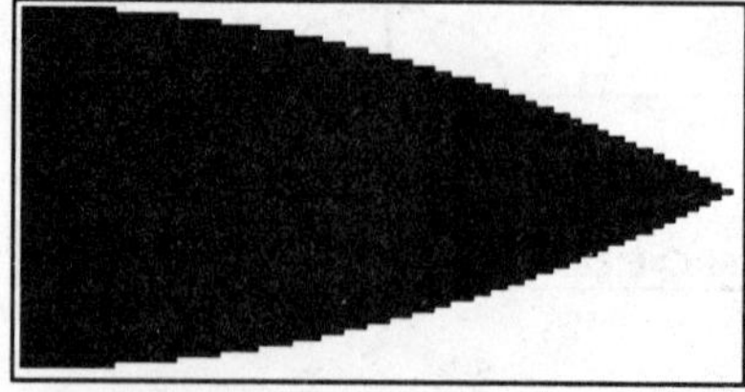

Two special cases of volume by cross-section is when a function $f(x)$ is revolved about either the x or y axis. In either case, the cross-sections are circles. As an example, let's consider the case when $f(x) = e^{-x}$ on the interval $[0, 1]$. By graphing $f(x)$ and its reflection across the x axis, $-f(x)$, we can get a visual idea of what the solid looks like in the case of revolving about the x axis. The region between the curves has been shaded with the command **Shade(y2, y1, 0, 1)**.

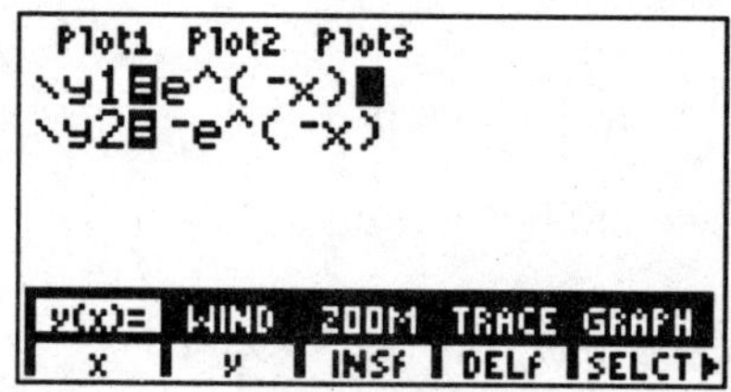
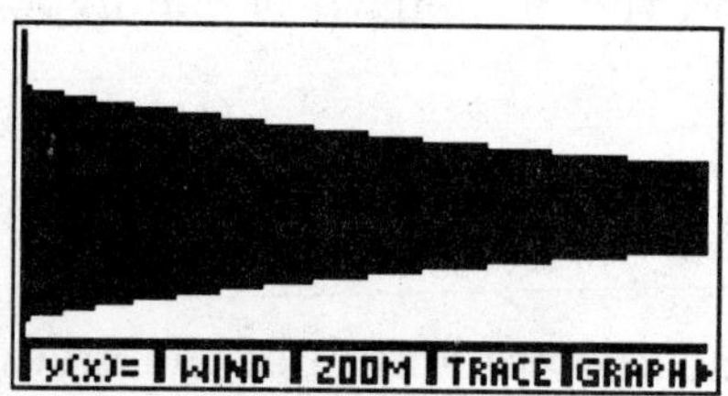

The volume can be calculated using the formula

$$V = \pi \int_a^b y(x)^2 \, dx$$

This can be numerically calculated using the **fnInt** command as shown.

For the volume obtained when $f(x)$ is revolved about the y axis, we can get an idea of what the object looks like by plotting $f(|x|)$ as shown in the screen shot. The volume has been shaded using the command

$$\text{Shade}(0, \ y1, \ -1, \ 1).$$

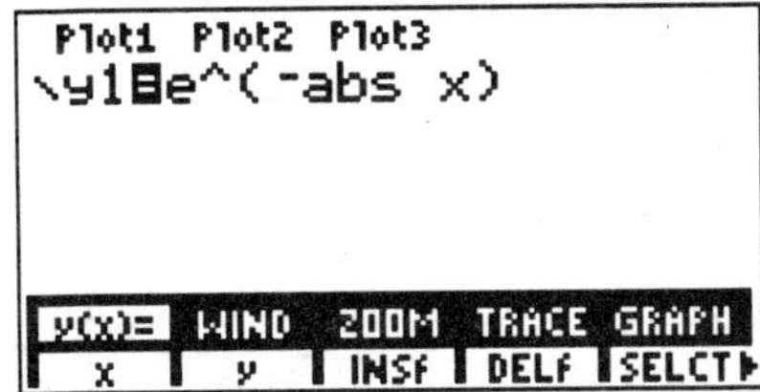 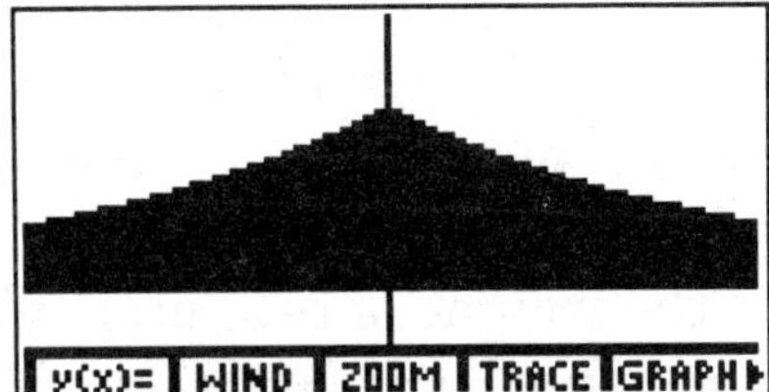

The volume can be calculated using the formula

$$V = 2\pi \int_a^b xy(x)\, dx$$

The numerical calculation of this integral is shown in a previous screen shot.

Often a volume is defined by a *region* in the plane which is rotated about one of the axes. For this type of problem, graphing the region can be very helpful in determining the integration limits. For example, if the region bounded by the curves $y = 2 - x^2$ and $y = e^x$ is rotated about the x-axis then a graphical solution is required to find the intersection of the two curves. The two curves are plotted below, and **ISECT** is used to find the intersections which correspond to the integration limits.

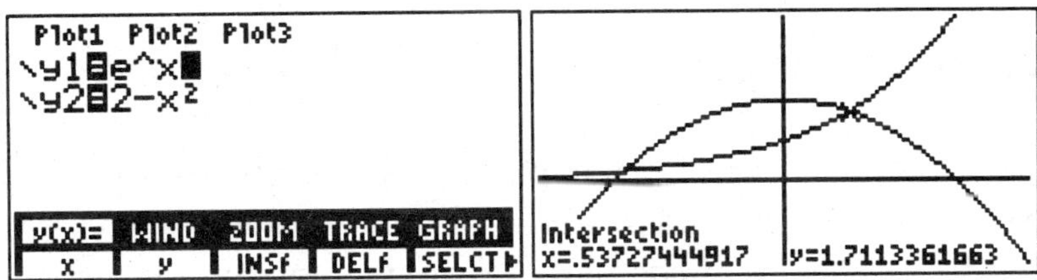

The x coordinate is saved in **ANS** and can be transferred to a different memory location so that it can used in the volume calculation as shown. The intersections become the limits of integration in this case.

```
Ans→A
              -1.3159737778
Ans→B
            .537274449174
π*fnInt((y2-y1)²,x,A,
B)
              4.28378777935
```

Exercises:

1. Find the volume of revolution about the x-axis for the region bounded by the curves $f(x) = -x^2 + 5x - 2$ and $g(x) = x$.

2. Find the volume of revolution about the y-axis for the region bounded by the curves $f(x) = -x^2 + 5x - 2$ and $g(x) = x$.

3. Find the volume of revolution about the x-axis for the region bounded by the curve $f(x) = 1/(1 + x^2)$, the lines $x = 0$ and $x = 1$, and the y-axis.

4. Find the volume of revolution about the y-axis for the region bounded above by the curves $y = 7\ln x$ and $y = 4 - x - x^3$, and below by the x-axis.

5. Find the volume of the solid obtained by revolving the region in exercise 4 about the x-axis.

6. Find the volume of the solid obtained by revolving the region in exercise 4 about the line $x = 4$.

6.4 Center of Mass

The balancing point $(\overline{x}, \overline{y})$ of a flat plate is called the center of mass. The coordinates of the center of mass can be found by calculating the ratio of the moment to the total mass of the plate. Formulas for the case when the plate has a uniform density, and the shape can be described by a function $f(x)$, are given by

$$\overline{x} = \frac{\int_a^b x f(x)\, dx}{\int_a^b f(x)\, dx} \qquad \overline{y} = \frac{\int_a^b \frac{1}{2} f(x)^2\, dx}{\int_a^b f(x)\, dx}.$$

Numerical calculation using the **fnInt** command allows one to find these quantities easily, regardless of the choice of $f(x)$.

As a simple example, let's find the center of mass of a uniform density region bounded by $y = \sqrt{x}$, the x-axis and the line $x = 1$. This region is shaded in the graph below. The density calculation follows.

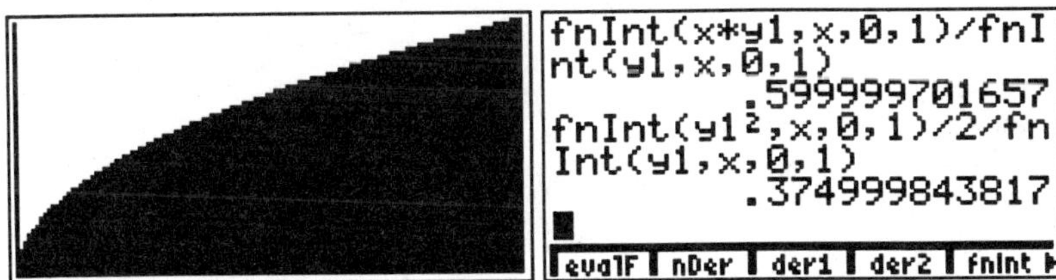

Note that the **fnInt** command can be used in algebraic operations.

Exercises:

1. Find the center of mass of the region bounded by the curves $f(x) = -x^2 + 5x - 2$ and $g(x) = x$.

2. Find the center of mass of the region bounded by the curve $f(x) = x^{1/3}$, the x-axis and the line $x = 8$.

3. Find the center of mass of the region bounded by the curves $f(x) = \sin x$, the x-axis and the line $x = \pi$.

Chapter 7

Differential Equations

In this chapter we examine how the TI-86 can be used in the study of differential equations. Once a differential equation is defined it can be solved numerically and the resulting graph can be drawn. The TI-86 has added features that allow the plotting of direction fields and the choice of numerical solution method. Related material can be found in Chapter 10 of Stewart's **Calculus**.

7.1   Solving a Differential Equation

The first step in solving any differential equation is to set the **GRAPH** mode for differential equations. This is done by pressing $\boxed{\textbf{2nd}}$ [**MODE**] and setting the fifth line so that **DifEq** is highlighted.

When the $\boxed{\textbf{GRAPH}}$ key is pressed, a special menu appears for the study of differential equations.

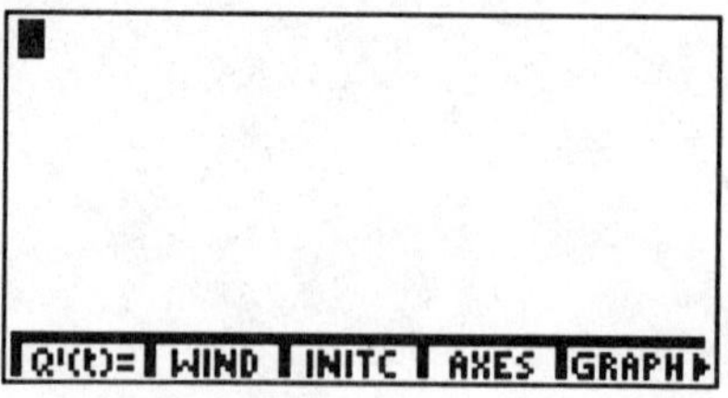

The differential equation is defined by pressing $\boxed{\textbf{F1}}$ for $\mathbf{Q'(t)} =$. Note that the independent variable is t, and the dependent variable is Qi, where i is an integer. So the differential equation must be in the form

$$Q_1'(t) = F(t, Q_1(t)).$$

The right hand side, F of the differential equation is entered in the same way that a function is entered. The explicit t dependence in Q_1 is not shown. For convenience, the letters t and Q are available by pressing $\boxed{\textbf{F1}}$ and $\boxed{\textbf{F2}}$ in the submenu of this command. For the logistic differential equation,

$$Q_1'(t) = Q_1 - Q_1^2$$

the screen looks like this:

You may define more differential equations by pressing $\boxed{\textbf{ENTER}}$ after each one.

The commands **INSf**, **DELf**, **SELCT**, **ALL+** and **ALL-** in this sub-menu, are used in the same way as they are for functions. **INSf** inserts a new equation, **DELf** deletes the highlighted equation, **SELCT** is used to select or deselect a differential equation to solve, **ALL+** selects all equations and **ALL-** deselects all equations. The one command that is different from those of the function graphing menu is the **STYLE** command, which is $\boxed{\textbf{F3}}$ after pressing $\boxed{\textbf{MORE}}$. This command sets the plotting style for the differential equation plot. There are four choices of plotting style:

- *thick lines*

- *thin lines*

- a *small circle* that traces out the solution either with the solution drawn or not drawn.

Pressing **STYLE** cycles through these choices. The type is indicated to the left of the equation. The *small circle* selection is shown in the screenshot below. The default setting is *thin lines*.

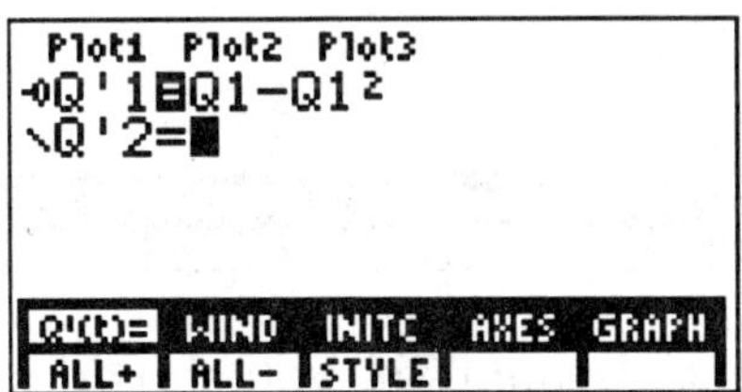

Once you have defined a differential equation, you need to set the window for the graph. There are several more settings here than in the standard function window settings. The new settings are **tMin, tMax, tStep, tPlot** and **difTol**.

The remaining settings, **xMin, xMax, xScl, yMin, yMax** and **yScl** are the standard ones that appear when plotting a function, and are defined in the same way by setting the dimensions of the window. For the single first order differential equation we are discussing, the x axis is identical to the t axis, so it makes sense to take **tMin = xMin** and **tMax = xMax**. The next value, **tStep**, is the increment in the t variable when the trace is used. The choice for **tStep** need not be small, since it does not effect the accuracy with which the numerical algorithm calculates the solution. In fact there is no use taking

$$\mathbf{tStep} < \frac{\mathbf{tMax} - \mathbf{tMin}}{126}$$

since only the plotting time is increased, and not the accuracy of the numerical solution. Note that the right-hand side of this inequality is just the spacing between pixels. **tPlot** is the value of t at which plotting begins. For most situations take **tPlot** to be the same as **tMin**. The last setting, **difTol** is a tolerance used in selecting the step size in the numerical computation. In most cases it can be left to its default setting of 0.001. A smaller value of this parameter will lead to greater accuracy, but it will also lead to slower calculations. The next item to set is the axes.

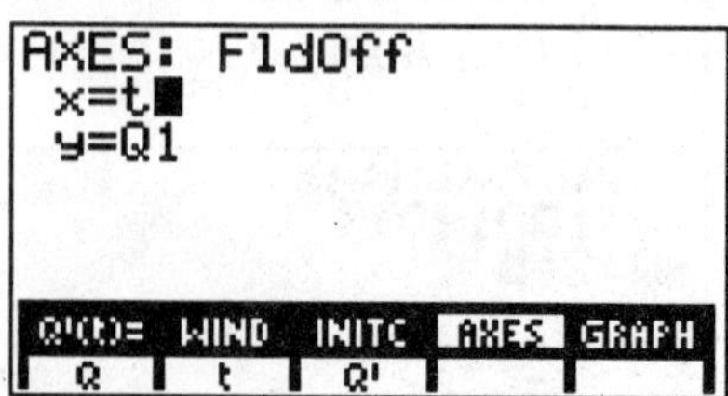

For a first order differential equation, you should set the x axis to be t and the y axis to be Q_1.

The solution to a differential equation, such as the logistic equation, is an infinite family of solutions that depend on a single arbitrary constant. Once you specify another piece of information, namely the value of Q_1 at a particular value of t, the solution is determined. This is called an *initial condition*, and takes the form

$$Q1(t_0) = Q1_0.$$

The initial condition is set by pressing $\boxed{\textbf{F3}}$ for **INITC** and entering the value of $Q1_0$ as shown.

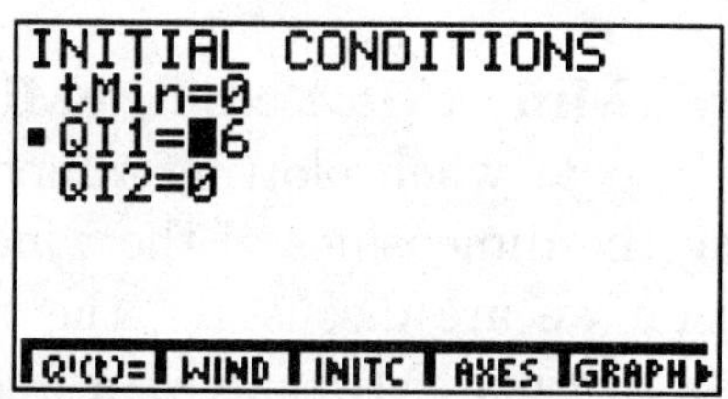

Note that t_0 is exactly **tMin** so this value is not needed. However, it can be modified in this screen instead of in the **WIND** screen. The bullet that appears to the left of QI1 indicates that a differential equation is defined corresponding to $Q1$. A bullet will appear for each differential equation that is defined, even though it may not be selected.

As for graphing solutions, the **FORMT** (format) command allows one to set various options like turning the axes on and off. It is accessed by pressing MORE and then F1 from the main differential equation graphing screen.

These options are set using the arrow keys and the ENTER key. The first four options are the same as those for graphing ordinary functions. **CoordOn** and **CoordOff** determine if the coordinates of the point are shown in the **TRACE** mode. **AxesOn** and **AxesOff** determine if the axes are drawn for a plot. **GridOn** and **GridOff** determine if grid points are shown on the graph or not. Grid points will only appear if you set both of **xScl** and **yScl** to a nonzero value. **LabelOn** and **LabelOff** is the option for having labels on the axes. The last two lines in the format options are for choosing a numerical method and choosing if a direction field is drawn. These will be discussed later in the chapter.

The graph of a solution to a differential equation is much different than the graph of a function. When you graph a function with your calculator you define a function by an explicit formula which the calculator uses to evaluate and then plot the points. With a differential equation, we only have an implicit formula relating the function and its derivative. The calculator uses a numerical scheme to approximate the solution to the differential equation and then plots these approximate values.

The numerical schemes used by the TI-86 are the Runge-Kutta algorithm and the Euler algorithm. To change numerical methods, enter the **FORMT** options screen. Use the arrow keys to move to the fifth line which sets the numerical method. Choose either **RK** for Runge-Kutta or **Euler** for the Euler method. The Runge-Kutta method is usually used since it gives greater accuracy. Once a numerical method is chosen, the solution can be calculated. To see the graph of the numerical solution to the differential equation, press $\boxed{\text{F5}}$ for **GRAPH**.

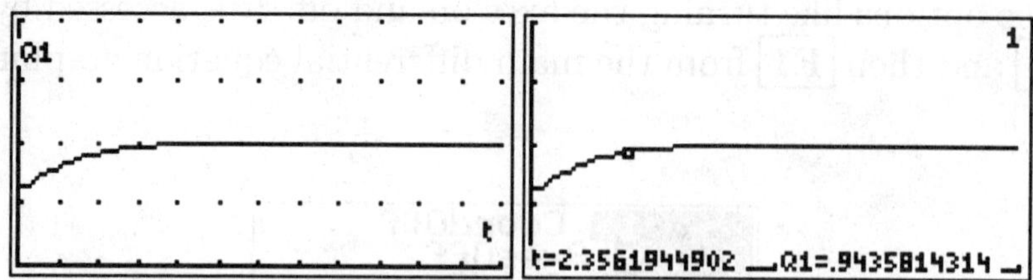

The screenshots show the graph with and without labels and gridpoints. Pressing $\boxed{\textbf{MORE}}$ in the main graphing window scrolls the menu through the fifteen choices. We discuss the important ones below.

TRACE can be used to find the coordinates of points on the graph just like for regular functions. The screen will show the value of t and $Q1$ as shown in the above screenshot. The various **ZOOM** commands are used in the same way as for functions to locate particular features of the graph. **ZIN** zooms in on a particular feature of the graph, **ZOUT** zooms out for a distant view, **BOX** zooms in on the region you choose by using the arrow keys and **ZFIT** sets the window so that the solution fits nicely within it. There are several other zoom commands as well as these important ones. The **EVAL** command is used to evaluate the solution at a particular value of t.

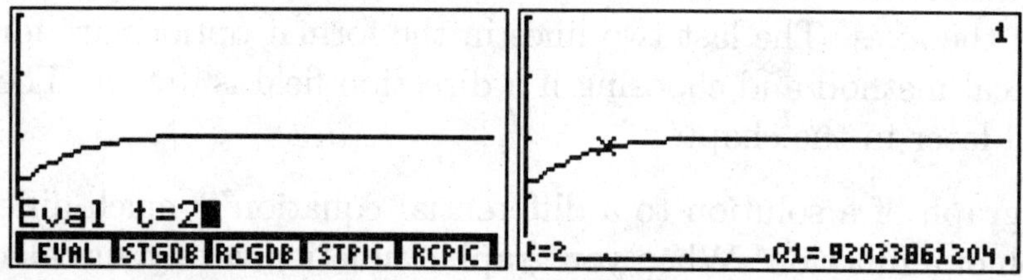

If you know the exact solution to the differential equation then you can compare the numerical solution with the exact solution by using the draw function command. To do this, press $\boxed{\textbf{MORE}}$ from the basic graph menu, $\boxed{\text{F2}}$ for **DRAW** and then $\boxed{\textbf{MORE}}$ again, followed by $\boxed{\text{F2}}$ for **DrawF** (draw function). You will then be prompted for a function to plot. You must enter a function of x even though the independent variable is t.

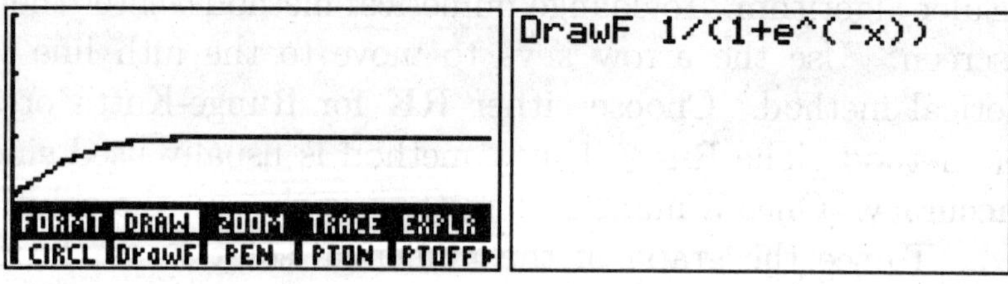

You can then compare the exact with the numerical approximation to the solution. You will generally find good agreement between exact solutions and those computed using the Runge-Kutta numerical method.

You may also use the **DrawF** command to plot other functions relevant to the differential equation besides the exact solution. The **EXPLR** (for explore) command is unique to the differential equation graphing screen. Press $\boxed{\textbf{MORE}}$ and then $\boxed{\textbf{F5}}$ and the cursor will appear at the center of the display. Use the arrow keys to move the cursor around to an initial condition you would like to see the solution to the differential equation plotted for. Then press $\boxed{\textbf{ENTER}}$ to start solving the equation and plotting the solution curve.

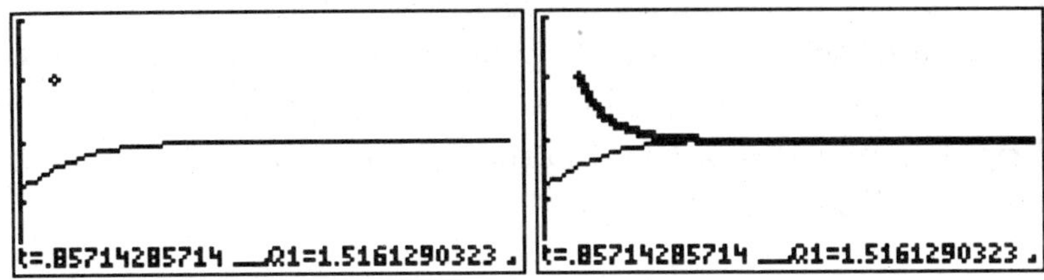

This procedure can be repeated so that you can examine several solutions to the differential equation for different initial conditions.

If the plot window is getting too cluttered with solution curves and functions, use the **CLDRW** command in the **DRAW** menu to clear the drawing and redraw the original solution that is determined by the differential equations selected and the initial conditions specified.

Exercises:

1. Show that $y = 2 + e^{-x^2}$ is a solution of the differential equation $y' + 2xy = 4x$. Compare this solution with the numerical solution calculated using the Runge-Kutta algorithm.

2. Show that every solution of the form $y = Ce^{-x^2/2}$ solves $y' = -xy$ where C is an arbitrary constant. Solve this equation numerically for the initial conditions $y(0) = 2$ and $y(0) = -3$. What is the connection between the initial conditions and the value of C?

3. Compare the Runge-Kutta and Euler numerical solutions to

$$y' = y\left(e^{2-y} - 1\right), \quad y(0) = 0.01$$

for $0 \le t \le 2$

4. Repeat exercise 3 for the initial value problem

$$y' = \sin(t^2 + y), \qquad y(0) = 0$$

for $0 \le t \le 5$.

5. Repeat exercise 3 for the initial value problem

$$y' = 0.5y \left(3 - 2\sin(2\pi t) - y\right), \qquad y(0) = 1$$

for $0 \le t \le 5$.

6. Check that the following functions are solutions of the given differential equation for each choice of C by defining the function as **y1** (for each choice of C) and then defining **y2** to be the difference between the left and right hand sides of the differential equation. In each case, the plot of **y2** should be zero.

   (a) $y = Ce^{-3t}, \quad y' = -3y, \quad C = \pm 1, \pm 2$

   (b) $y = \frac{2}{1 + Ce^{-2t}}, \quad y' - 2y = -y^2, \quad C = -9, -0.5, 0, 3$

   (c) $y = \frac{(1+3t)^{4/3} + C}{(1+3t)^{1/3}}, \quad y' + \frac{y}{1+3t} = 4, \quad C = 0, 25, 50$

7. Solve $y' = y \sin x$ exactly and numerically for the initial condition $y(0) = 1$. Is the solution a periodic function of x? If so, what is the period?

7.2   Systems and Higher Order Equations

The TI-86 calculator can be used to solve systems of first order differential equations. In addition, since higher order differential equations can be transformed to a system of first order differential equations, the TI-86 can also be used to solve higher order differential equations.

Consider the general second order differential equation

$$y''(t) = F(t, y(t), y'(t))$$

where F is a given function. Introduce the auxiliary variables, $Q1(t)$ and $Q2(t)$ where $y(t) = Q1(t)$ and $y'(t) = Q2(t)$. Substituting these variables into the original differential equation gives

$$
\begin{aligned}
Q1'(t) &= Q2(t) \\
Q2'(t) &= F(t, Q1(t), Q2(t))
\end{aligned}
$$

which is a system of first order equations.

For example, consider

$$y''(t) = -y(t)$$

The process above leads to the first order system of differential equations given below:

$$
\begin{aligned}
Q1'(t) &= Q2(t) \\
Q2'(t) &= -Q1(t)
\end{aligned}
$$

These equations are easily entered in the $\mathbf{Q'(t)} =$ window.

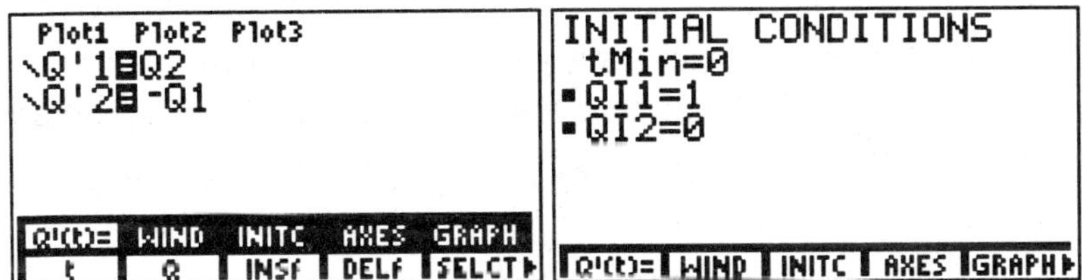

You must also set the initial conditions before you plot. In this case you must give values for $Q1$ and $Q2$ at the starting value **tMin**. These correspond to the function and derivative values at **tMin** for $y(t) = Q1(t)$.

There are several ways to display the solution to this equation. You can plot $Q1$ vs. t, $Q2$ vs. t, or $Q2$ vs. $Q1$. The latter plot is called the *phase plane plot*. Your choice is set from the **AXES** command.

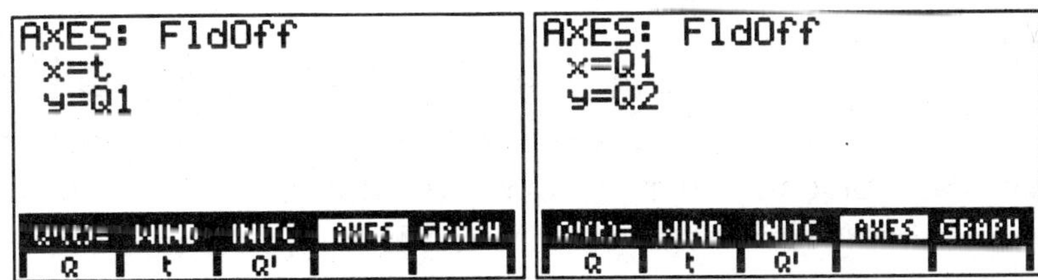

In the window settings **WIND**, it is no longer the case that the x and t settings are the same. Set **tMin** to be the value of t where the initial conditions are given and choose **tMax** to be about 5 to start. If you are unsure of the x and y dimensions, choose something reasonable to start, and then use the zoom to fit (**ZFIT**) command to get a better setting. The next screen shot shows a plot of the solution in the phase plane. Note that in the first screen the trajectory appears to be incomplete. When this is the case, increase the value of **tMax** and replot the solution to see if there is any change. In this example there is.

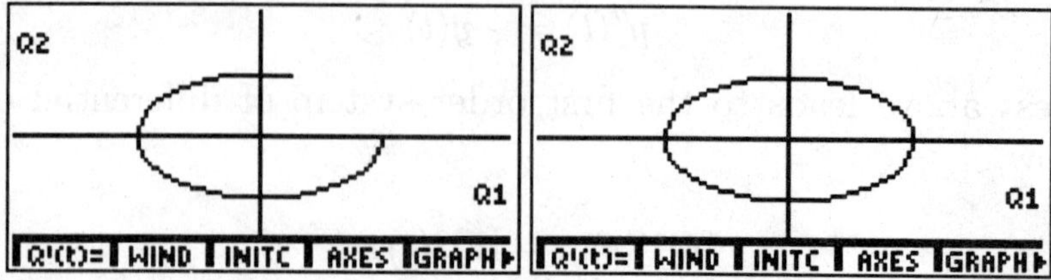

The technique of reducing one second order differential equation to two first order equations can be generalized to higher order equations. An Nth order equation can be reduced to N first order equations by introducing new variables $Q1, Q2 \ldots QN$ and defining them to be the function and its $N - 1$ derivatives. This can lead to quite a few choices for viewing the solution.

Exercises:

1. Write the differential equation

$$\frac{d^2 x}{dt^2} + \omega^2 x = 0$$

   as the system of two differential equations

   $$\frac{dx}{dt} = y \qquad \frac{dy}{dt} = -\omega^2 x$$

   by first defining $y = dx/dt$. Use the TI-86 to calculate the solution corresponding to $\omega = 2$, $x(0) = 1$ and $x'(0) = -2$, and view it as a phase plane plot in the x-y plane. What conic section do the curves correspond to? Verify this by showing that $y(t)^2 + \omega^2 x(t)^2 =$ constant.

2. Write the differential equation

$$\frac{d^2 x}{dt^2} - \beta^2 x = 0$$

as the system of two first order differential equations

$$\frac{dx}{dt} = y \qquad \frac{dy}{dt} = \beta^2 x.$$

by first defining $y = dx/dt$. Use the TI-86 to calculate the solution corresponding to $\beta = 2$, $x(0) = 1$ and $x'(0) = -2$, and view it as a phase plane plot in the x-y plane. What conic section do the curves correspond to? Verify this by showing that $y(t)^2 - \beta^2 x(t)^2 =$ constant.

3. Verify that $x = \sin(2t) + \cos(2t)$ and $y = \sin(2t) - 3\cos(2t)$ satisfy

$$x' = x + 5y, \qquad y' = -x - y$$

Plot this solution for $0 \le t \le \pi$.

4. Verify that $x = -e^t \sin(2t)$ and $y = e^t \cos(2t)$ satisfy

$$x' = x - 2y, \qquad y' = 2x + y$$

Plot this solution for $0 \le t \le \pi$.

5. Compare the Runge-Kutta and Euler methods for approximating the solution to

$$x' = y, \quad y' = -x - y^3$$
$$x(0) = 5, \quad y(0) = -3$$

for $0 \le t \le 2$.

7.3 Direction Fields

The TI-86 can be used to draw the direction field for a given differential equation or a first order system of 2 differential equations.

To show the direction field for a differential equation, first set the sixth line in the format options. There are three choices, **SlpFld** for slope field, **DirFld** for direction field and **FldOff** for field off. The **SlpFld** command is used when working with a single first order differential equation. The **DirFld** command is used for working with systems of first order differential equations. Last of all, if you don't want to see the direction field, turn it off with the **FldOff** setting.

Once the direction field has been turned on, proceed in the usual way to defined the differential equation, the window, the axes and the initial condition(s). The next screen shot shows the direction field for

$$Q1' = Q1 - t^2$$

on a window of size $-4 \leq t, x \leq 4, -4 \leq y \leq 4$. Short lines are drawn on a equally spaced grid that indicate the slope of the tangent line to the solution at that grid point. The number of lines drawn for the direction field is determined by setting the variable **fldRes** (the field resolution). This variable can be set from the **AXES** command. A larger value of **fldRes** means more lines are drawn for the direction field. The choice of number determines how many direction field lines you get horizontally, so in the example there are 15 sets of lines. The thicker line indicates a particular solution.

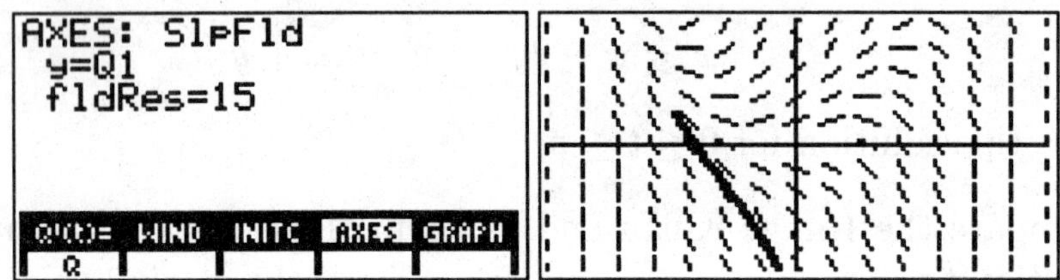

The command **EXPLR** can be used to draw several solution curves on the direction field. From the main graphing menu press $\boxed{\text{MORE}}$ and then $\boxed{\text{F5}}$ for **EXPLR**. A cursor will appear at the screen center with the coordinates shown. Use the arrow keys to move the cursor around and then press $\boxed{\text{ENTER}}$. The solution curve that passes through the chosen point, which is a new initial condition, is then drawn. This procedure can be repeated for as many initial conditions as you like. The result is shown in the next screenshot.

For a system of first order equations $Q1' = Q2, Q2' = -Q1$, the format setting of **DirFld** can be used to draw direction fields. The variable **fldRes** again sets the number of lines drawn. Explore is used in the same way to draw several solution curves. The screen shot shows the results for this system of equations.

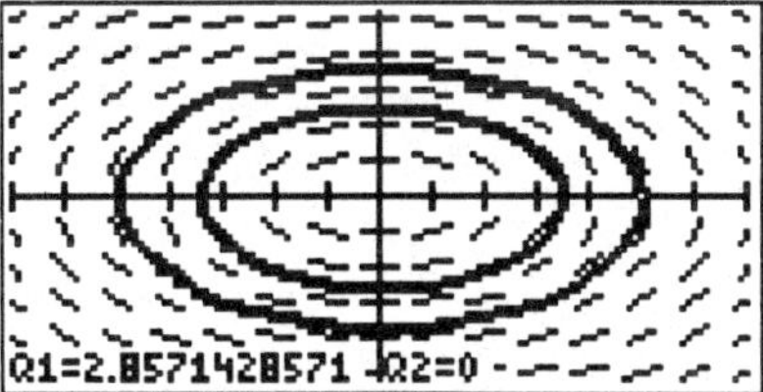

For systems of differential equations with more than two unknowns, you cannot use the direction field. In fact you will get an error message if you try and set the axes to anything but $Q1$ and $Q2$. In this latter case, you will get a direction field plot, but it is meaningless. In fact, you will notice that solution curves are no longer tangent to the direction field.

Exercises:

1. Plot the direction field by hand for the differential equation $y' = x - y$. Draw several solutions based on your direction field. Compare your answers with the numerical evaluated solutions and direction fields computed by the TI-86.

2. A particular population of two species, where one species preys on the other, is modeled by the Lotka-Volterra equations

$$\frac{dP}{dt} = 3P - PQ$$
$$\frac{dQ}{dt} = -Q + 0.5PQ$$

where $P(t)$ is the prey population and $Q(t)$ is the predator population for $t \geq 0$. Plot solutions and direction fields for the Lotka-Volterra equations for several initial conditions in the P-Q plane and the P, Q versus t plane. What can you say about the behavior of the two populations? Does either population become extinct?

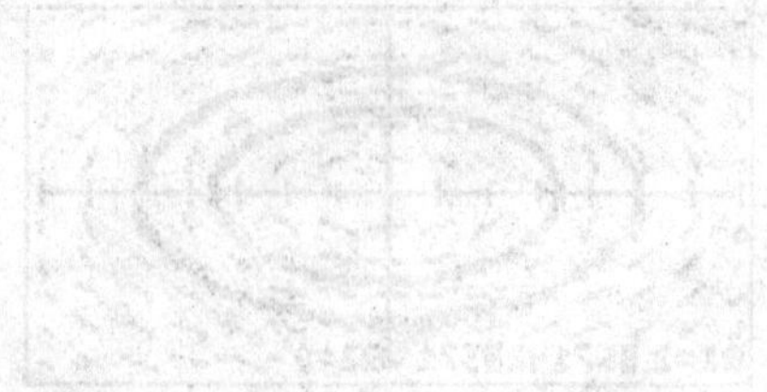

Chapter 8

Sequences and Series

In this chapter we see how the TI-86 calculator can be used to explore sequences and series. The TI-86 has built-in commands for displaying sequences and sums. In addition, the **seq** and **sum** commands can be used to analyze Taylor Polynomial approximations of functions. Related material can be found in Chapter 12 of Stewart's **Calculus**.

8.1 Sequences

A finite subset of a sequence is represented on your calculator as a ordered list. A list looks like

$$\{1, 2, 5, 7, 9\}$$

where the first element is 1, the second is 2 and so on. A sequence on the other hand is an infinite set of numbers defined by some rule. For example

$$a_n = \frac{1}{2^n} \qquad \left\{\frac{1}{2}, \frac{1}{4}, \frac{1}{8}, \dots\right\}.$$

Of course, the TI-86 cannot view the entire sequence, but it can view any finite subset that has consecutive elements. There are two ways to do this. One way is to press $\boxed{\textbf{2nd}}$ [**MATH**], $\boxed{\textbf{F5}}$ for **MISC** and then $\boxed{\textbf{F3}}$ for **seq** (which of course stands for sequence). The other is to press $\boxed{\textbf{2nd}}$ [**LIST**],

$\boxed{\textbf{F5}}$ for **OPS** (operations), and then $\boxed{\textbf{MORE}}$ followed by $\boxed{\textbf{F3}}$ for **seq**.

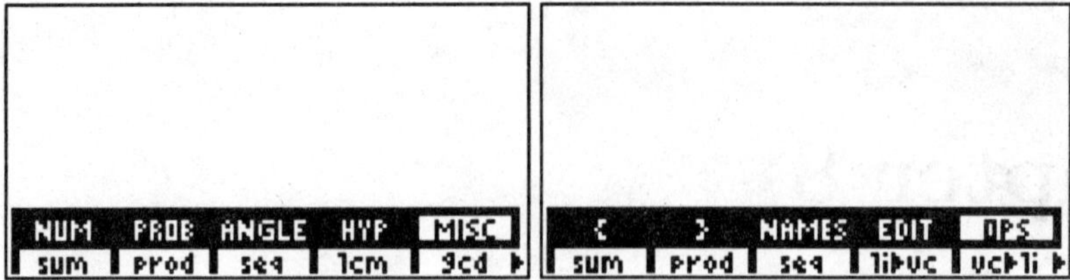

The correct syntax for the **seq** command is

$$\textbf{seq}(\textbf{a}(\textbf{n}), \textbf{n}, \textbf{p}, \textbf{q}, \textbf{r})$$

where $a(n)$ is the sequence defined relation for the terms, n is the index variable, p is the starting value of the index, q is the last value of the index and r is the increment size. For most sequences studied in calculus, r is taken to be 1. The starting and ending values need not be integers for this command, but they will be for the type of sequences studied in calculus. The screen shot shown illustrates how to display the sequence corresponding to

$$a_n = \frac{1}{2^n}$$

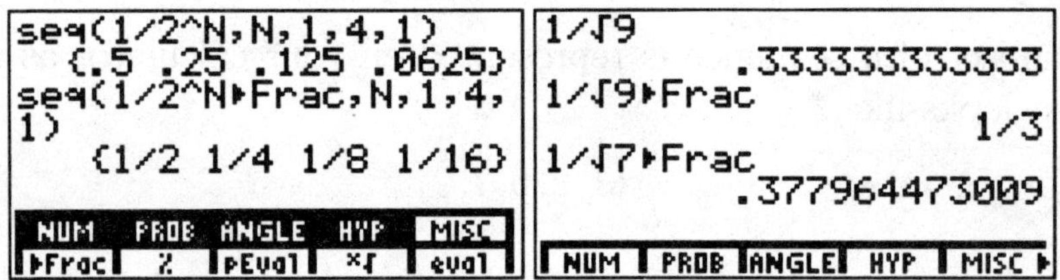

The second sequence shows how you can display the results as a fraction instead of a decimal using the ▶ **Frac** command. The ▶ **Frac** command will express a decimal as a fraction when possible. This command is part of the **MATH** menu. It is found by pressing $\boxed{\textbf{F5}}$ for **MISC**, $\boxed{\textbf{MORE}}$ and $\boxed{\textbf{F1}}$. Note that the number will only be converted to a fraction when it is possible.

If you set the index parameters so that more terms are calculated than can viewed on the screen, then the left and right arrow keys can be used to scroll through the terms. The sequence elements calculated with the **seq** command can be saved in memory by pressing $\boxed{\textbf{STO▶}}$ and then the name of the variable. For example, you can call the sequence A and then a given element can be recalled using the notation $A(i)$, where i is an integer.

```
seq((-1)^N/N,N,1,100,    seq(1/ln N,N,2,50,1)→
1)→A                     B
{-1 .5 -.33333333333…    {1.44269504089 .9102…
A(20)                    B(1)
              .05                   1.44269504089
A(75)                    B(2)
       -.013333333333             .910239226627
```

Note in the second screen above that although the starting value of the index is 2, the first element of B is indexed starting at 1. This is always the case when storing a sequence in memory since a list is indexed starting at 1.

The important quantity that is calculated with sequences is the limit

$$\lim_{n \to \infty} a_n$$

Use the sequence command to calculate the sequence values for large values of n. This can give an indication of what the limit may be. You might even set the index increment to be a large value so that you can examine terms far out in the sequence. For example, if

$$a_n = \frac{n}{n+1}$$

then the terms for $n = 500, 1000, 1500$ and 2000 are calculated using an increment of 500.

```
seq(N/(N+1)▶Frac,N,50
0,2000,500)
{500/501 1000/1001 1…


  NUM   PROB  ANGLE  HYP   MISC
 ▶Frac   %   ▶Eval   ×√   eval
```

The values indicate that the limit is 1, but this must be shown, since we have only looked at four sequence terms.

A graph of a sequence can be plotted that allows one to view 126 terms all at once. In the standard function graphing window, set the window so that the spacing between pixels is one. For example take $1 \le x \le 127$ or any interval of size 126. Choose the y range values to view the terms in a reasonable way. Define a function to be the continuous version of the sequence term definition. For example

$$a_n = \frac{n}{n+1} \iff f(x) = \frac{x}{x+1}$$

and graph the result. It is best that the graph is done in the **DrawDot** mode since a sequence is a discrete function defined only at the integers.. The resulting graph is shown in the next screenshot.

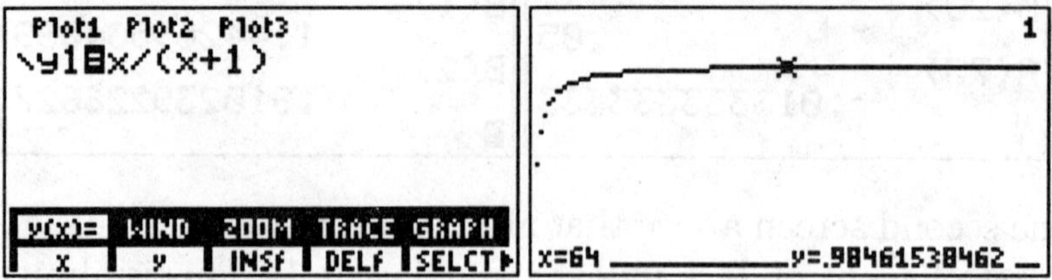

When you use the **TRACE** function, the y-coordinate of the cursor is always the sequence term corresponding to an index equal to the x-coordinate. So in the screen shot, the value of a_{64} is shown. The limiting behavior can clearly be seen in this example.

If the limiting behavior is not clear from the graph then choose a different interval that is also of length 126. For example $1000 \le x \le 1126$ will show the sequence farther out for larger index values.

A graph of a sequence that diverges will be quite clear as you will see no accumulation of data points. For example, consider the sequence $a_n = \sin(n)$. The plot appears to be a series of dots randomly placed on the screen.

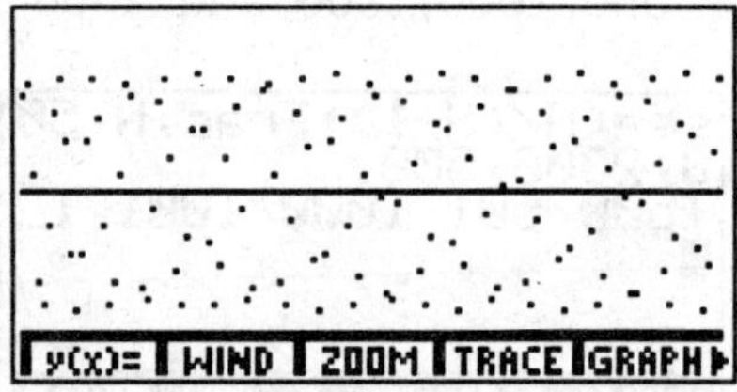

A plot is also helpful in deciding if a sequence is increasing or decreasing, or if it is bounded above or below. Since you can only look at a small window of the whole sequence, this is not a proof that is increasing (decreasing) or bounded. But it does suggest an hypothesis that can be tested.

Exercises: For questions 1-6, plot the sequence and determine whether or not the sequence has a limit.

1. $a_n = \dfrac{n^2 - 15n + 50}{n^2 + 5n + 6}$

2. $a_n = \dfrac{\ln n}{n^{1/4}}$

3. $a_n = \frac{(-1)^n n^2 + 4}{n^2 + 19n}$

4. $a_n = \left(1 + \frac{2}{n}\right)^{3n}$

5. $a_n = \frac{n^3 + 15n^2 + 4}{n^3 + 1}$

6. $a_n = \sqrt{\cos n}$

8.2 Infinite Series

An infinite series is of the form

$$\sum_{k=1}^{\infty} a_k$$

Series are usually studied by considering the sequence of partial sums defined by

$$S_n = \sum_{k=1}^{n} a_k$$

Partial sums are easy to evaluate by combining the **sum** command with the **seq** command. The **sum** command can be accessed from either the **OPS** set of commands in the **LIST** menu or the **MISC** set of commands in the **MATH** menu. The syntax for this command is

$$\textbf{sum seq}(\textbf{a}(\textbf{k}), \textbf{k}, \textbf{p}, \textbf{q}, \textbf{r})$$

where $a(k)$ gives the general form of the term a_k, k matches the index for the term, p is the starting value of the summation index, q is the final value of the summation index and r is the summation index increment (which is usually 1). The series above have $p - 1$, but some series may start at 0, 2 or perhaps another integer. To evaluate the partial sum S_{100} for the two infinite series

$$\sum_{k=1}^{\infty} \frac{1}{k^2} \quad \text{and} \quad \sum_{k=1}^{\infty} \frac{(-1)^k}{k^2}$$

enter the commands shown in the following screen shot.

```
sum seq(1/K²,K,1,100,
1)
           1.63498390018
sum seq((-1)^K/K²,K,1
,100,1)
          -.822417533374
■
NUM | PROB |ANGLE| HYP | MISC ▶
```

The $\boxed{\text{2nd}}$ [ENTRY] command can be used to repeat the long command and calculate the partial sum for a different number of terms. If you choose a large value for the final summation index, you may find yourself taking a coffee break while waiting for your calculator to complete this task. If you find that a calculation is taking a long time, just press the $\boxed{\text{ON}}$ button to stop the calculation.

For an infinite series the important quantity is the sum of the series. That is, the limiting behavior of the sequence of partial sums S_n as n tends to infinity. The TI-86 can be used to predict convergence and estimate the sum by using the **sum seq** command. As an example consider the infinite series

$$\sum_{k=1}^{\infty} \frac{1}{2^k}.$$

and do repeated partial sums that contain a successively higher number of terms. The screenshot illustrates the process for calculating S_{10}, S_{20} and S_{30}.

```
sum seq(1/2^K,K,1,10,
1)
            .9990234375
sum seq(1/2^K,K,1,20,
1)
          .999999046326
sum seq(1/2^K,K,1,30,
1)
```

The values of the sums suggest that the limit is 1, and in fact this is easily checked as this series is a geometric series. This procedure has some drawbacks. The first is that it often takes a very large number of terms in the partial sum to get a reasonable estimate of the sum. Some series converge slower than others. This may mean you have to wait a long time for the approximate sum to be calculated. For most convergent geometric series, the convergence is fast, in the sense that a small number of terms in the partial sum gives a fairly accurate value of the exact sum. The second drawback is that this procedure is very inefficient due to the fact that each time you calculate a new partial sum you are adding up a lot of the same numbers again. To calculate S_{20}, S_{10} is recalculated and the next ten terms are added to it.

A nice solution to the problem of finding an approximate sum is to write a program that efficiently calculates the partial sums given the terms in the series. The algorithm is extremely simple and an implementation of this is given in chapter 9. Assuming the a_k are known

$$a_1 \;\rightarrow\; S_1$$
$$a_n + S_{n-1} \;\rightarrow\; S_n$$

gives the partial sums for $n = 1$ to some final value N. This has the added benefit that you have the first N terms of the sequence of partial sums, and these can be graphed just like any other sequence. Skip ahead to the next chapter and read how to enter a program into your calculator and then enter the series program. The program automatically sets up the graphing screen to have the two sequences. A is the list corresponding to the sequence elements and S is the sequence of partial sums. Both of these sequences are plotted for the first 126 terms. The **TRACE** or **EVAL** functions can be used to find approximate partial sums.

When this approach is applied to $\sum_{k=1}^{\infty} 1/2^k = 1$, the convergence is clearly illustrated. The first screenshot shows how to enter the information for the program.

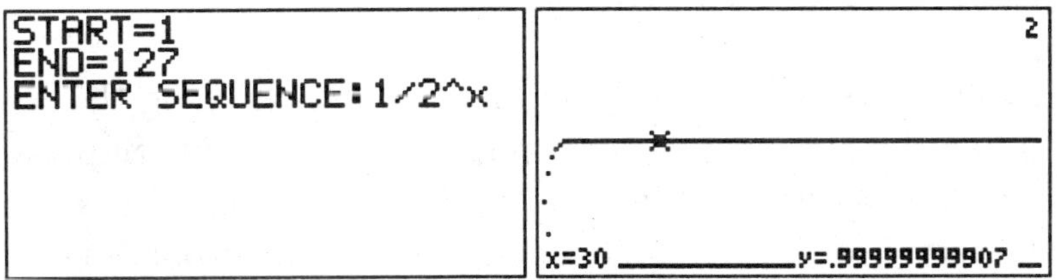

For the divergent harmonic series, $\sum_{k=1}^{\infty} 1/k$, a graph suggests the slow logarithmic growth for the series (plot $\ln x$ on the same graph and compare).

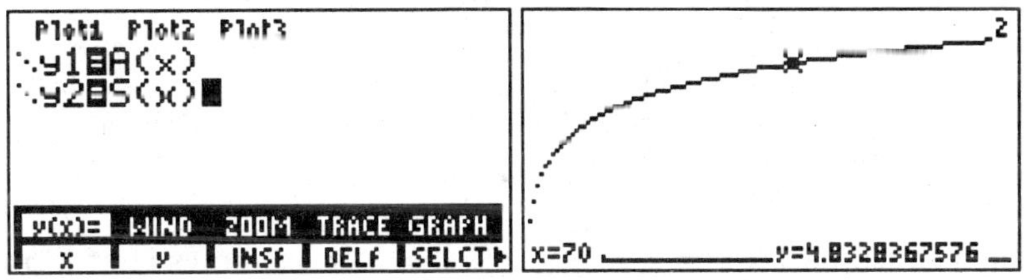

On the same graph, you can also plot the terms in the series. This will serve as a check on the limiting behavior of the a_k. If the graph doesn't show the terms approaching zero then the series must diverge.

Exercises: For questions 1-6, plot the sequence of terms and the sequence of partial sums for each of the series. (To plot the partial sums use the program given in Chapter 9). Use your results to conjecture whether the given series converges or diverges. In the case of convergence, give an approximate value for the sum.

1. $\sum_{k=1}^{\infty} \frac{1}{k}$

2. $\sum_{k=1}^{\infty} \frac{1}{k^4}$

3. $\sum_{k=1}^{\infty} \frac{(-1)^k}{k!}$

4. $\sum_{k=1}^{\infty} \frac{1}{k}$

5. $\sum_{k=1}^{\infty} \frac{k^3}{3^k}$

6. $\sum_{k=1}^{\infty} \frac{\sin k}{k^2}$

8.3   Convergence Tests

A graphing calculator can be used as a first step in checking the comparison test for deciding if a given series is convergent or divergent. Suppose you are comparing the two series, $\sum a_k$ and $\sum b_k$. Using a range of $1 \le k \le 127$ (or some other interval of the same length, graph the corresponding continuous functions to compare their behaviors. Remember that you must know the convergence properties of one of these two series when using this test.

As an example, let's apply this idea to the two series

$$\sum_{k=1}^{\infty} \frac{1}{k} \quad \text{and} \quad \sum_{k=1}^{\infty} \frac{\ln k}{k}.$$

Graphing the functions $f(x) = 1/x$ and $g(x) = \ln x/x$ shows that $g(x) > f(x)$

provided that $x > 3$.

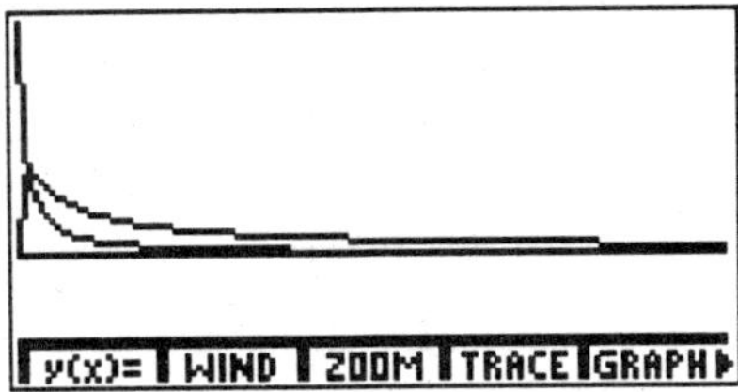

The inequality $x > 3$ is found by using the trace is locate where the intersection of the two curves is. Remember that the inequality has only been checked on a finite interval and it must be true for all values of x. But this piece of information gives you confidence that the inequality is correct and you can attempt to check it for all values of x. Consequently, since the harmonic series $\sum_{k=1}^{\infty} \frac{1}{k}$ diverges, we expect that the series $\sum_{k=1}^{\infty} \frac{\ln k}{k}$ also diverges.

Alternating series $\sum_{k=1}^{\infty} (-1)^k b_k$ are tested for convergence by examining the limiting behavior of the b_k and deciding if they decrease to zero. A graph of the corresponding continuous function or the sequence values will suggest the limiting behavior. **TRACE** can be used in deciding if the sequence is decreasing. To test the series

$$\sum_{k=1}^{\infty} \frac{(-1)^{k+1}}{k + \sqrt{k}}$$

for convergence, plot a graph of $f(x) = 1/(x + \sqrt{x})$. It appears from the resulting graph that the terms decrease to zero. It is important to remember that just because it is true on this interval it doesn't make it true for all values of k. The general case must be checked by examining how successive terms compare. An alternative to using **TRACE** to see if the sequence decreases with k, is to plot the derivative of $f(x)$ using the built-in derivative function **der1**.

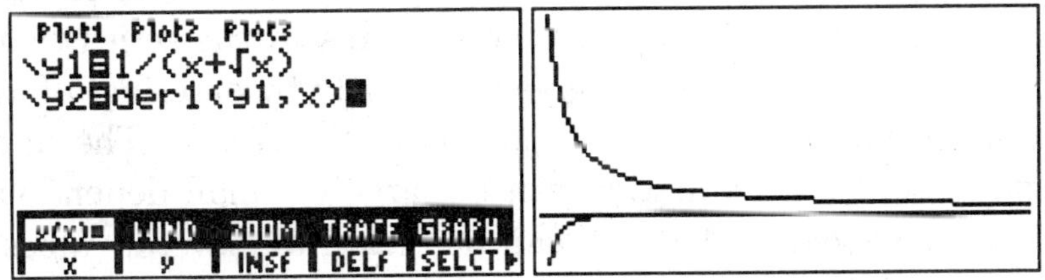

If the derivative is negative, then the elements are decreasing.

The techniques discussed earlier in this chapter about estimating the sum of the series apply to alternating series as well. The following screenshot shows the graph of the partial sums and the terms in the alternating series including their sign.

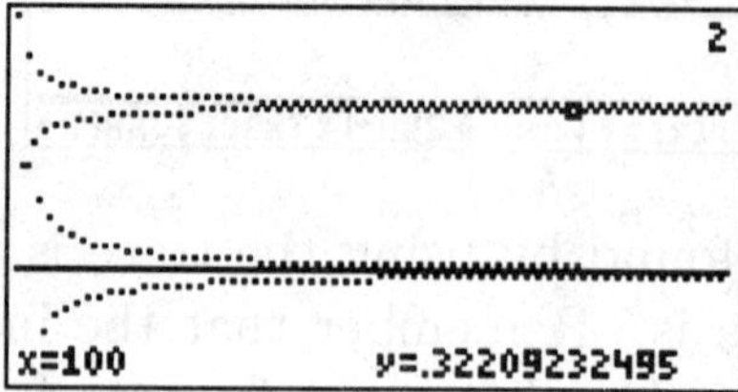

For this series the sum is approximately 0.32.

Exercises: Apply a convergence test to determine the convergence or divergence of the series given in the exercise set of the previous section.

8.4   Power Series

A power series is an expression of the form

$$\sum_{k=0}^{\infty} a_k x^k$$

Power series are typically used to approximate functions. For this reason, it is important to determine the values of x for which the series converges. We are not able to directly graph power series, but we can plot the partial sums. By taking

$$\mathbf{y1 = sum\ seq(a(k)\ x\char94 k, k, 0, N, 1)}$$

as the function definition for **y1** (or any other) where $a(k)$ is the coefficient term in the series and N is the number of terms in the series that will be used. The coefficient terms can be given explicitly by entering a formula for $a(k)$ or by using **seq** to evaluate the terms in the series. The upper limit N in the finite sum approximation can be large or small depending on the nature of the convergence. A graph can be drawn for several values of N to see the progress of the convergence. Of course, larger values of N will take a lot of time to plot.

As a simple example take $a_k = 1$ for all k. The resulting series is the geometric series

$$\sum_{k=0}^{\infty} x^k$$

which converges only when $|x| < 1$. In fact, this geometric series converges to $1/(1-x)$. To investigate this example using the TI-86, take

$$\mathbf{y1 = sum\ seq(x\hat{\ }k, k, 0, N, 1)}$$

where N will start with the value 10. The resulting graph for $-1 < x < 1$ compares favorably with the known sum for the series. Note that the series doesn't do well near $x = \pm 1$, but this is expected since the series doesn't converge at these points.

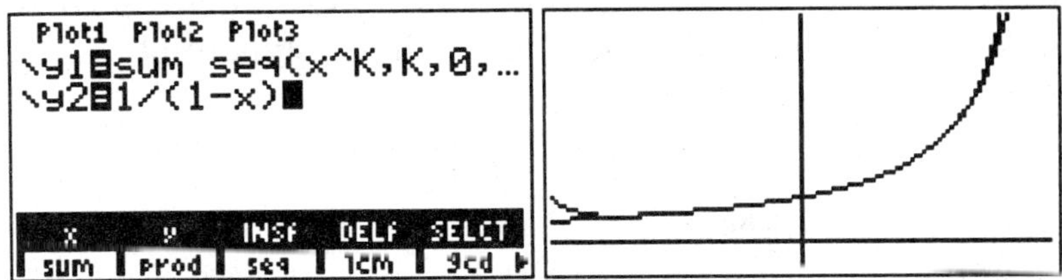

By taking a larger value of N the convergence for x near ± 1 (with $|x| < 1$) will become better.

The truncation of a power series is an example of a MacLaurin polynomial. By plotting graphs of various MacLaurin polynomials we can get an idea about the accuracy of such an approximation. For example, consider the series

$$\sum_{k=0}^{\infty} \frac{(-1)^k}{(2k+1)!} x^{2k+1}$$

which converges to $\sin x$ for all x. The factorial

$$k! = \begin{cases} 1, & k = 0 \\ 1 \cdot 2 \cdot 3 \cdots k, & k = 1, 2, 3, \ldots \end{cases}$$

is defined only for integers and is found by pressing $\boxed{\textbf{2nd}}$ [MATH], $\boxed{\textbf{F2}}$ for **PROB** (which stands for probability) and then $\boxed{\textbf{F1}}$ for the factorial command. Enter

$$\mathbf{y1 = sum\ seq((-1)\hat{\ }K * x\hat{\ }(2K+1)/(2K+1)!, k, 0, N, 1)}$$

and start with $N = 3$. This is the third degree MacLaurin polynomial. The screenshots for $N = 3$ and 5 graphed against $\sin(x)$ can be used to find the values of x for which the approximation is good.

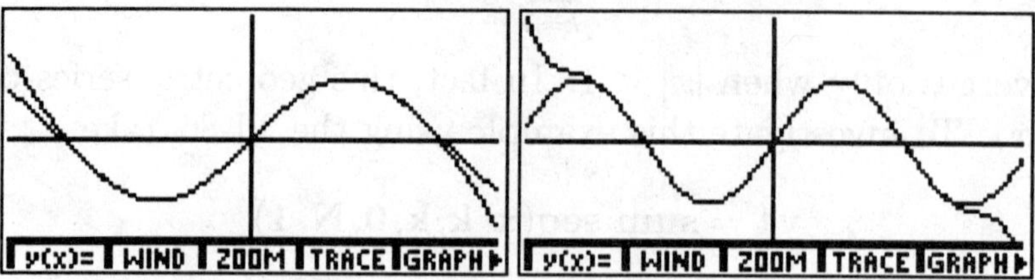

The advantage of entering a general expression for the MacLaurin series is that it is very easy to change the order of the MacLaurin polynomial by editing the function definition and modifying only the value of N. If you are only interested in a particular order and don't expect to look at more than one or two polynomials, then just enter that polynomial. For the $\sin(x)$ example enter

$$\mathbf{y1 = x - x^3/6}$$

instead of the general expression.

Taylor polynomials are just as easy to work with. Simply replace x^k by $(x - a)^k$ in the function definition, where a is the point that is expanded about.

As a final note about Taylor and MacLaurin polynomials, the coefficients are not always given by simple definitions like those in $\sin(x)$ or e^x. When this is the case, you should forgo the general case and work with individual polynomials. It won't be quite so easy to change the order of the polynomial as you will have to compute the new coefficient each time and edit the polynomial in the function screen to incorporate the changes.

Exercises: Plot the partial sums of the power series in exercises 1-5 to determine the interval of convergence.

1. $\sum_{k=0}^{\infty} 2^k x^k$

2. $\sum_{k=0}^{\infty} \frac{(-1)^k x^{2k}}{(2k)!}$

3. $\sum_{k=1}^{\infty} \frac{kx^k}{3^k}$

4. $\sum_{k=1}^{\infty} \frac{(-1)^{k+1} x^k}{k}$

5. $\sum_{k=1}^{\infty} \frac{(x-1)^k}{k(k+1)}$

6. Give the 8th degree Taylor polynomial approximation of $\cos(x)$ centered at $x = 0$. Plot the difference between the Taylor polynomial and $\cos(x)$ on the interval $-\pi \leq x \leq \pi$ and determine the maximum error associated with this approximation. Compare this error with the value from the remainder term for Taylor polynomials given in Stewart's **Calculus**.

7. Use the fourth degree Taylor polynomial for $\sqrt{1 + x^4}$ centered at $x = 0$ to approximate the integral

$$\int_0^1 \sqrt{1 + x^4}\, dx$$

Estimate the error in the approximation.

Chapter 9

Programming

This chapter serves as an introduction to the programming capabilities of the TI-86. We describe how a program is entered, edited and executed. We also discuss the basic input and output commands which allow numbers and functions to be entered, and results to be displayed. Control commands are used to determine the direction and order in which commands are executed. Several example programs are given which are useful in the study of Calculus. These also serve to illustrate how commands are used to achieve different results. Readers who are not interesting in programming should at least read the first section.

9.1 Entering, Editing and Executing a Program

There are two ways to enter a program into the TI-86. One is use the link cable to download a program from someone else's calculator. This is clearly the simplest method, since someone else has done all the work. Once the program is in the calculator's memory, all that's left to do is run it. The other approach is to enter the commands keystroke by keystroke. To do this press the $\boxed{\text{PRGM}}$ button. The following screen shot shows the two commands available: **NAMES** and **EDIT**.

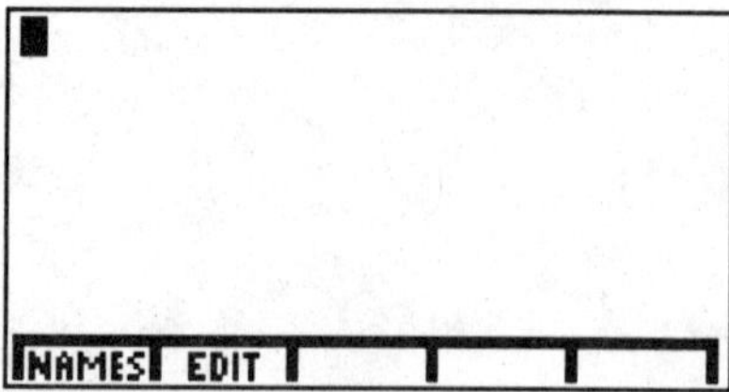

The next step is to press $\boxed{\textbf{F2}}$ (for **EDIT**) to bring up the next screen shot. You have a choice of entering a new name at the alphabetic cursor if you are entering a new program. If you have already entered a program, you may want to edit the program. In this case you choose one of the programs by pressing the corresponding function key. In the screen shot there are four existing programs to choose from for editing.

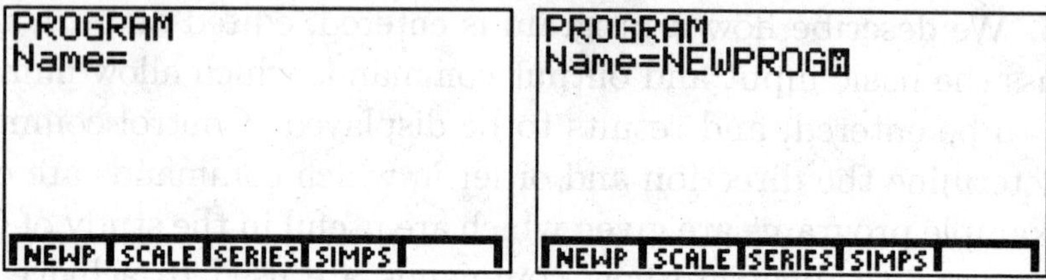

Program names can be up to eight characters long, but only four to six characters will show up in the menu. Once you have entered a name for the program, you can enter the first program line at the colon. Each program line is separated by a colon. The screen shot shows the menu choices when entering or editing a program.

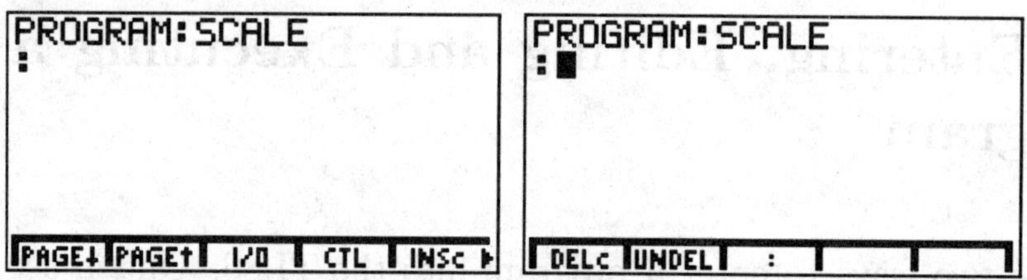

The first two menu choices are **PAGE↓** and **PAGE↑** which allow you to scroll through a long program quickly. The up and down arrow keys can also be used to scroll through single lines of the program. The next two commands, **I/O** and **CTL**, open up submenus; one for input and output commands, and the other for control commands. The next three commands are useful for editing. **INSc** will insert a command above the current line where the cursor is currently located. **DELc** is for deleting the current program line that the cursor is on and **UNDEL** will undelete the last command deleted if you realize you don't really want to delete the line. As always, the insert and delete keys can be used to modify a particular command and the arrow keys can be used to move around in the program. Each time you press $\boxed{\textbf{ENTER}}$ the cursor moves to a new line and puts a colon to separate the new program line from the previous. It is best for editing if each line corresponds to a single command. But if a command is short you many want to have more than one command on a single line. In this case, use the : menu item ($\boxed{\textbf{MORE}}$ and $\boxed{\textbf{F3}}$). The colon character is also on the shift decimal point key.

After you have finished editing or entering a program press $\boxed{\textbf{EXIT}}$ to leave this mode. The screen shot shows how the program **SCALE** will look like after it has been entered in memory.

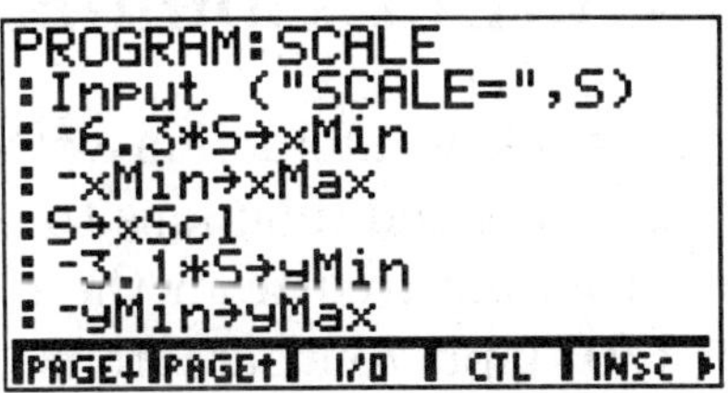

Once a program is in memory, it can be executed by pressing $\boxed{\textbf{PRGM}}$, $\boxed{\textbf{F1}}$ for **NAMES**, and then the function key corresponding to the program you want to run. If your program is taking an unusually long time to run you may have a problem with a particular part of the program. You can stop the execution by pressing $\boxed{\textbf{ON}}$. The second screen below shows the result of stopping the program. Notice that a **BREAK** error message appears. Either press $\boxed{\textbf{F5}}$ for **QUIT** to end program execution, or $\boxed{\textbf{F1}}$ for **GOTO** to see the line at which you stopped the program execution. This can be helpful for trouble shooting a program, since this line may be causing a problem.

If you find you no longer want a particular program, it can erased by pressing $\boxed{\textbf{2nd}}$ [MEM]. In this menu choose **DELET** for deleting programs and other variables. Choose the function key corresponding to **PRGM** (MORE and F5) and then a list of your currently stored programs will appear. Use the cursor to select the one you want to delete and then press ENTER to complete.

WARNING!!! There is no command to un-delete a program.

9.2 Input and Output Commands

Input and output commands allow interaction with the keyboard. These types of commands allow numbers and functions to be entered at a program prompt, and information to be received from the display. To access these commands, press F3 for **I/O** (input/output) from the programming menu. There are thirteen commands in this menu on the TI86. The first ten are shown in the screen shots below.

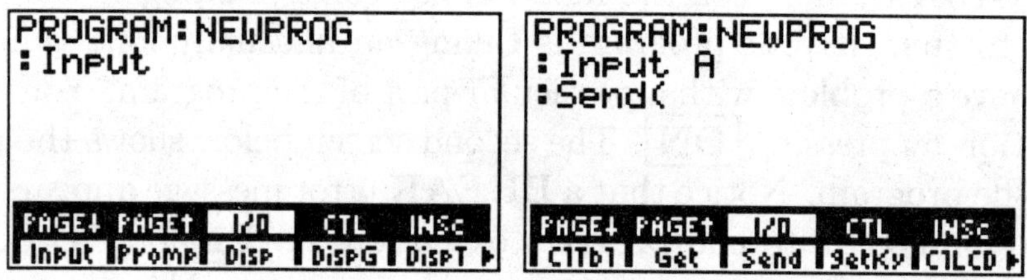

In this section we will only discuss the most important ones that are basic to programming. Discussion of the others can be found in the guidebook that comes with your calculator.

The commands **Input** (F1), **Prompt** (F2), and **InpSt** (MORE twice and then F3) are used for loading numbers and functions into the program. **Input** can be used in several ways. If no arguments are added to this command then the graph screen is displayed with the current graph and the cursor is shown with its coordinates. This allows the user to give coordinates to the program from the **GRAPH** screen. If an argument is given to the **Input** command then the user is prompted for a numerical value. For example, the command **Input A**, when running in a program, expects you to enter a value for A.

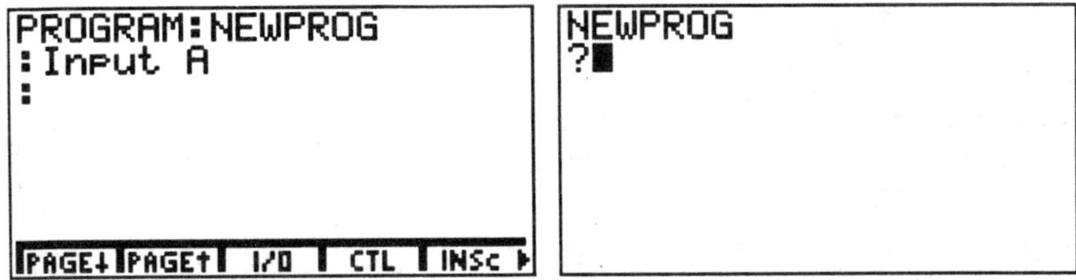

This is not the best way to use this command as you have no idea what variable is being requested, unless of course you are very familiar with the program. It is better programming style to instead write

$$\mathbf{Input}(\text{``}\mathbf{A}=\text{''}, \mathbf{A})$$

which will display

$$\mathbf{A} =$$

when executed. The quotation marks are also part of the **I/O** command set.

The **Prompt** command is similar to the **Input** command as it will allow you to enter numerical values when executed. It is especially good for entering several variables in succession. As an example

$$\mathbf{Prompt\ B, C, D}$$

asks for values of these three variables with an appropriate prompt each time. The results of using **Input** and **Prompt** are shown in the screens below.

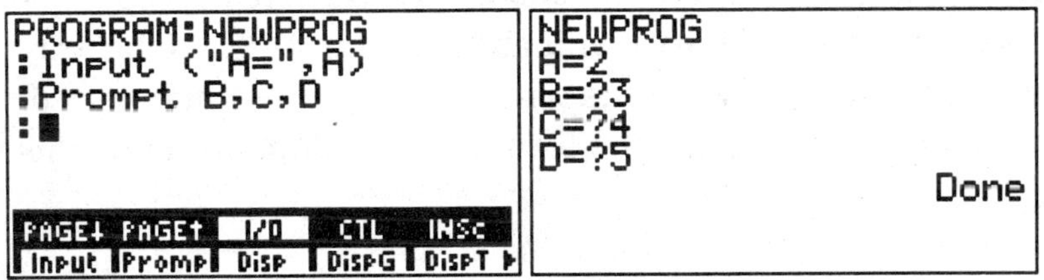

Often in a program, it is convenient to request the input of a function. This is done using the **InpSt** command (input a string). A string is just a list of alphanumeric characters and symbols. When the calculator receives a series of keystrokes, it doesn't think of them as a function. However, this can

be converted to a function using the **St▶Eq** (string to equation) command.
This command is accessed from the **String** menu. Press $\boxed{\text{2nd}}$ [**STRNG**]
and then $\boxed{\text{F5}}$ to find this command.

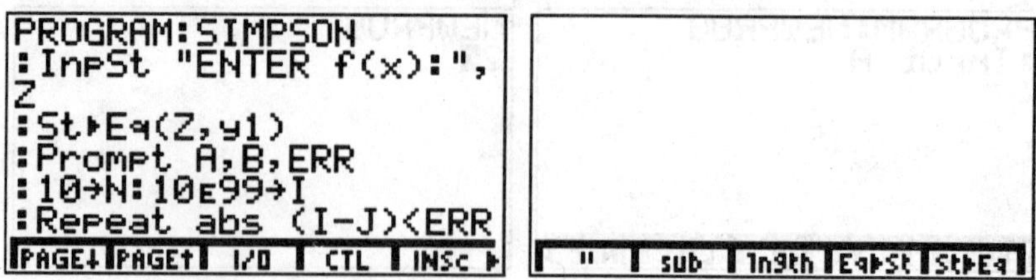

The correct sequence of commands for entering a function is shown in the
screen shot for the program that implement's the Simpson's numerical approximation to a definite integral (see the last section in this chapter). Note
that the formula entered from the keyboard for $f(x)$ will be stored in the
variable Z. The string to equation command converts Z into the function **y1**.
Of course you can use other variable names in place of Z or **y1**.

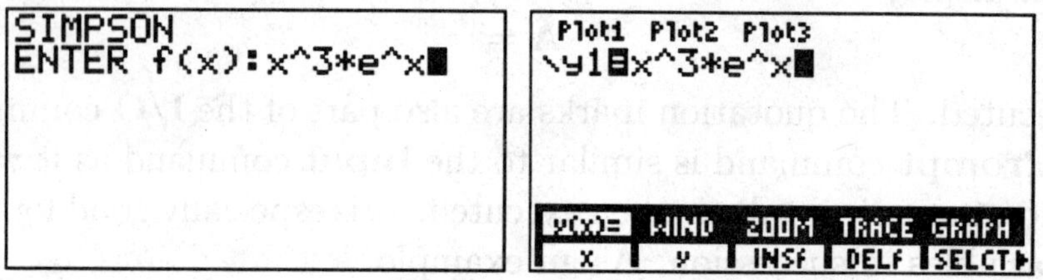

The screen shots show that you enter the string in the same way that you enter any function. The string to equation conversion gives the correct function
in the graphing screen.

The commands **Disp** ($\boxed{\text{F3}}$) and **DispG** ($\boxed{\text{F4}}$) are used to display text
or numbers on the home screen and the **GRAPH** screen respectively. There
are three ways that the **Disp** command can be used. One way is use it
without any arguments. The result is to display the home screen. Another
way is to include arguments in the form of variable names. In this case, the
syntax might be

$$\textbf{Disp A}, \textbf{B}, \textbf{C}$$

and the result is the numerical value of the variables A, B, and C displayed
on consecutive lines of the home screen. You can also add a caption to a
display by using quotation marks to set off the text. For example

$$\textbf{Disp ``The answer is''}, \textbf{A}$$

gives the second screen below.

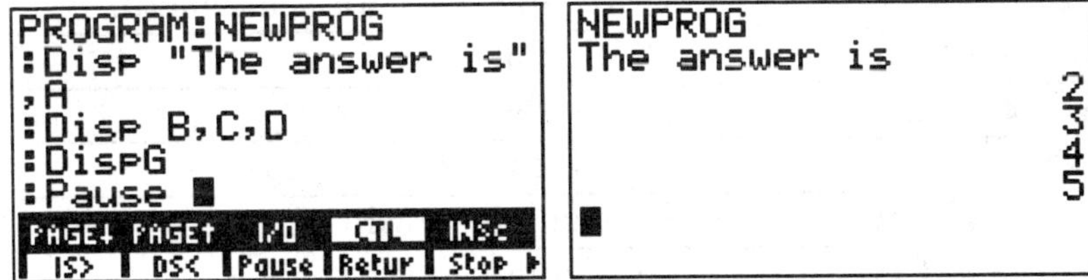

The **DispG** command takes no arguments and displays the current graph screen. If any function is selected then the graph will be drawn. If there are no functions selected then an empty graph will be shown.

The **PAUSE** command is often the next command after a display command, since it will temporarily halt the program. This gives you a chance to look at the results and make decisions about the future course of the program. Pressing ENTER restarts execution of the program. If a display command is the last command of a program there is no need to include the **PAUSE** command. The **PAUSE** command is part of the **CTL** menu for controlling the flow of a program.

Another command in the **I/O** menu related to the display is the clear display command, abbreviated **ClLCD** (MORE F5). It is sometimes nice to see a fresh clear home screen when a program first starts execution. In this case, the **ClLCD** command will be the first line of the program. This command can also be used at a later stage in the program when new input is going to occur.

9.3 Simple Scaling Program

In this example program, you will see how some of the simple **I/O** commands can be combined to create a time saving useful program.

It has already been noted that a nice choice of window is one with dimensions $-6.3 < x < 6.3, -3.1 < y < 3.1$. This type of window has the distinction that the spacing between points on a graph is exactly 0.1. Multiples of this window are also useful since the spacing between points can be set to any nice number we like. The program **SCALE** asks the user to enter a scaling factor (a positive number). The program sets the new window to be $-6.3S < x < 6.3S, -3.1S < y < 3.1S$ where S is the scale factor. The value of Δx is also displayed.

:Input ("SCALE=",S)	Prompts user for scale factor S
:-6.3*S▸xMin	Calculated xMin
:-xMin▸xMax	Calculates xMax
:S▸xScl	Defines xScl
:-3.1*S▸yMin	Calculates yMin
:-yMin▸yMax	Calculates yMax
:S▸yScl	Defines yScl
:Disp "$dx=$"	Display on screen $dx=$
:0.1*S	Calculates Δx and writes to the screen

The first line prompts the user for a value of S to be used as a scaling factor.
For example, entering the value 2 results in a spacing of 0.2 between pixels.
The next six lines modify the range settings for the graph using the store
command. Next the **Disp** line writes the new value of dx to the display.

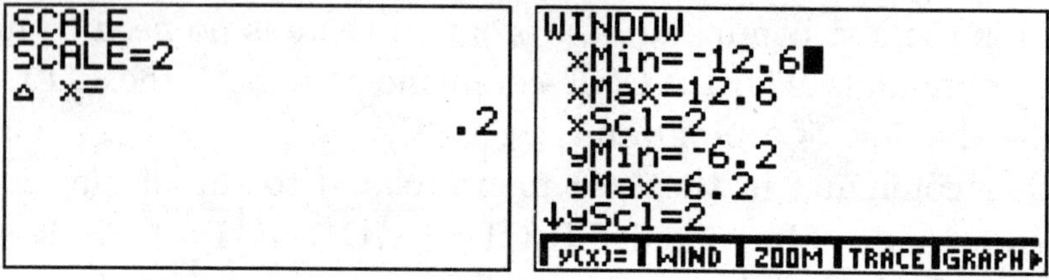

9.4   Control Commands

Control commands allow the programmer to determine the flow of a program
by directing it through various commands. The **CTL** menu is part of the
program editing screen and is accessed by pressing $\boxed{\text{F4}}$. There are eighteen
commands in this menu item on the TI-86. The first ten are shown in the
screen shots.

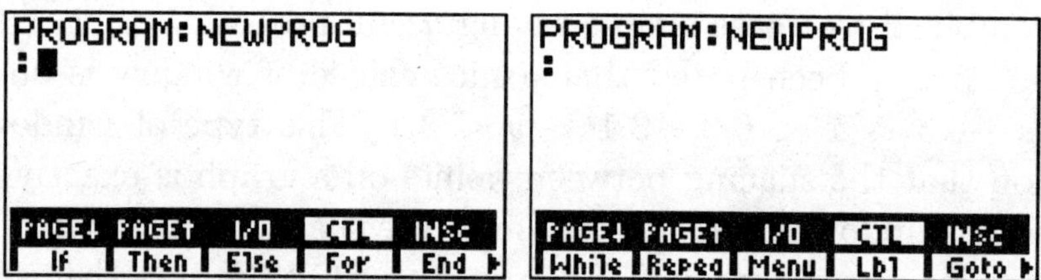

Again, we will only discuss the most important commands that are basic to
programming. Discussion of the others can be found in the guide book that
comes with the TI-86.

The commands **If** ($\boxed{\text{F1}}$), **Then** ($\boxed{\text{F2}}$), **Else** ($\boxed{\text{F3}}$) and **End** ($\boxed{\text{F5}}$) can be used when certain commands should be executed only when a particular condition is true. The condition should be one that is either true or false. Examples of this are $x > 2$ or $x = 3$. There are three ways these commands can be put together. The first way is to only use the **If** command. In this case the set of commands takes the form

> : **If** condition
>
> : command A
>
> : command B

with command A being executed if the condition is true, and skipped if the condition is false. Command B will always be executed regardless of the logical value of the condition. The following screen shows a typical example.

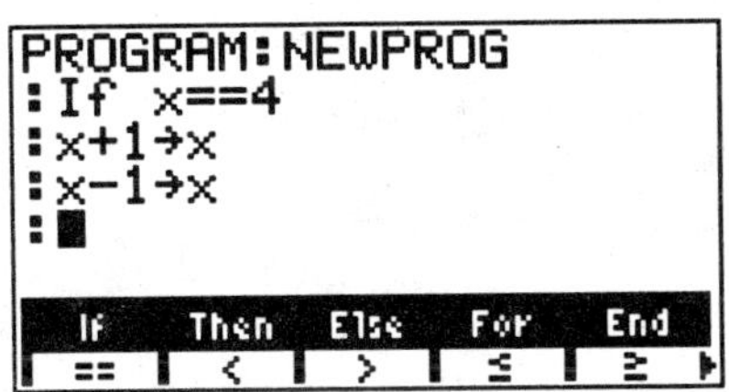

In the example, if $x = 4$ then x is incremented by one, and if $x \neq 4$ then x is decreased by one. Notice that the **TEST** commands are used when comparing quantities. These are accessed by pressing $\boxed{\text{2nd}}$ [**TEST**].

An **If** statement can be nested so that several conditions can be checked sequentially. If you want more than one command to be executed when the condition is true, then the commands **Then** and **End** should be added in the following way.

> : **If** condition
>
> : **Then**
>
> : commands
>
> : **End**
>
> : Other commands

The commands following the **Then** statement are only run when the condition is true. The **End** statement identifies the end of this set of commands.

The commands following the **End** are always executed. The screen below shows a simple example.

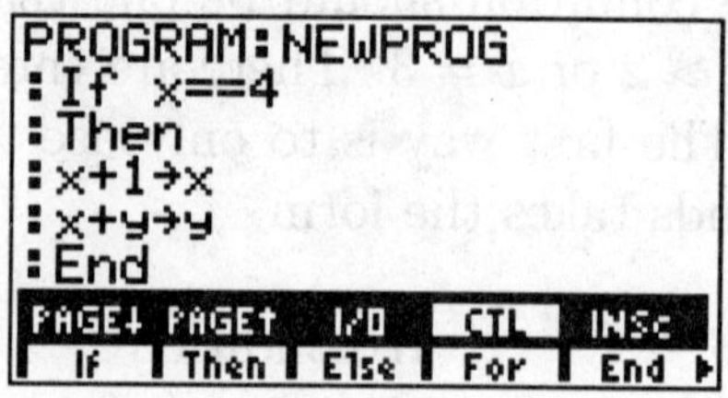

In the example, if $x = 4$, then x is incremented by one, and the result is added to y and stored in y. If $x \neq 4$, neither of these commands is executed.

Lastly, if you want one set of commands to be executed when the condition is true and another set of commands to be executed if the condition is false, then the **Else** command must be used. The format for an **If-Then-Else-End** structure is shown below.

> : **If** condition
>
> : **Then**
>
> : commands
>
> : **Else**
>
> : more commands
>
> : **End**
>
> : Other commands

If the condition is true, the commands following the **Then** command are executed, and if the condition is false, the commands following the **Else** command are executed. The other commands following the **End** command are always executed. In the example below, the variable x is checked to see if it is larger than 2. If this is true, then 4 is added to x and the value is stored in x. Otherwise 4 is subtracted from x and the value is stored in x. All commands after the **End** command are executed.

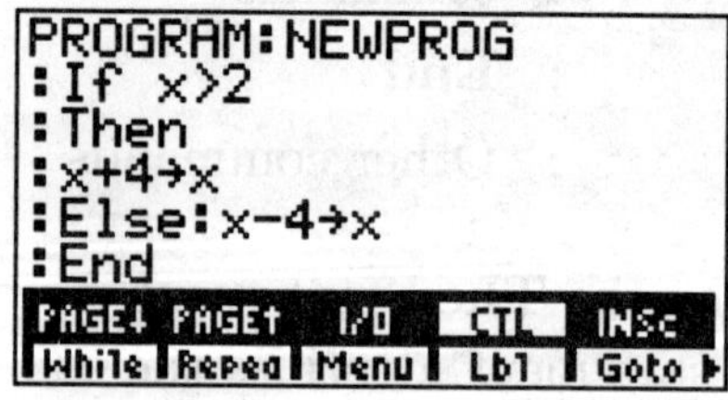

The **For** ([F4]) command is used to create a loop in a program. A loop is used to execute a set of commands a fixed number of times before moving on to the next command. The syntax for a **For** loop is

$$\textbf{For } (A,I,J,K)$$

where A is any variable name. The other parameters tell the starting value, the final value and the increment value respectively. K need not be an integer, but there should not be significant round off error so that J can not be obtained from I by increments of K. When this happens, the program will not work properly. The command should take the form

> : **For** (A,I,J,K)
> : Commands
> : **End**

The commands between the **For** and the **End** commands are done as long as I does not exceed J. When I exceeds J, these commands are no longer executed and the command after the **End** command is executed. The screen shot shows a simple example of a **For** loop.

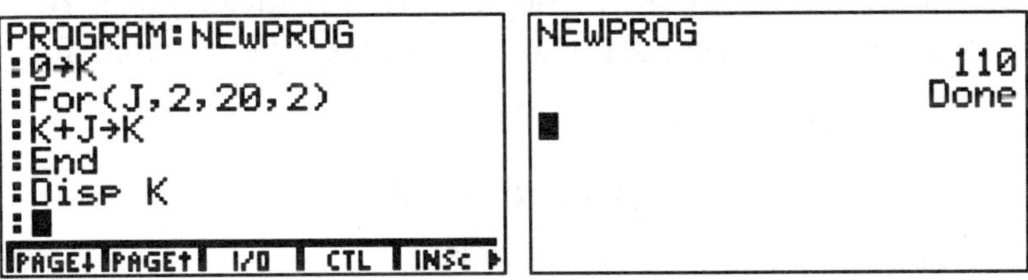

First the variable K is set to zero. Then the loop causes all even integers from 2 to 20 to be added to K, with the result displayed.

The **While** ([MORE] [F1]) command is used to execute several commands as long as a logical condition is true. This is different from the **If-Then** command where the commands are only done once if the condition is true. This means you must be careful that the commands executed within the **While** loop eventually cause the condition to become false. Otherwise, the program will run indefinitely. If the condition is false the commands will never be executed unless the program is redirected to the **While** loop from some other place in the program. An **End** statement is used to indicate the boundary of the **While** loop. The format for a **While** loop is

> : **While** condition

    : commands

    : **End**

    : Other commands

An example of a **While** loop is given in the screen below.

In this example, the variable A is initially set to zero. In the loop, one is added to its value as long as its value is less than ten. Note that the condition is eventually false, and the loop ends when A equals ten.

The **Repeat** command ($\boxed{\textbf{MORE}}\;\boxed{\textbf{F2}}$) creates a loop which is similar to a **While** loop, except the commands are always executed at least once since the condition is checked after the commands are executed. For a **While** loop the commands are not executed if the condition is false. For a **Repeat** loop, the commands are executed until the test is true. The format for a **Repeat** loop is

    : **Repeat** condition

    : commands

    : **End**

    : Other commands

An example of a **Repeat** loop is given in the screen below.

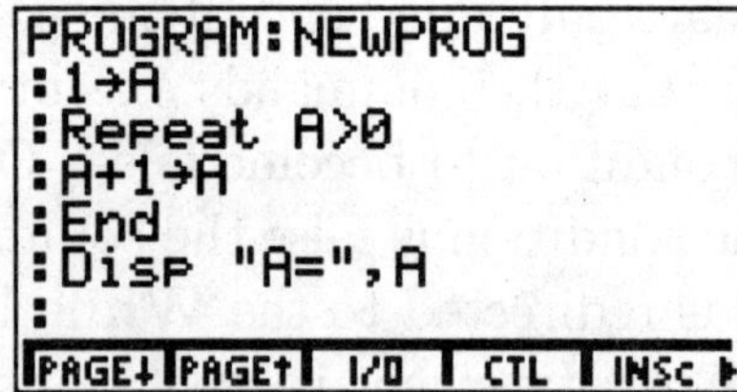
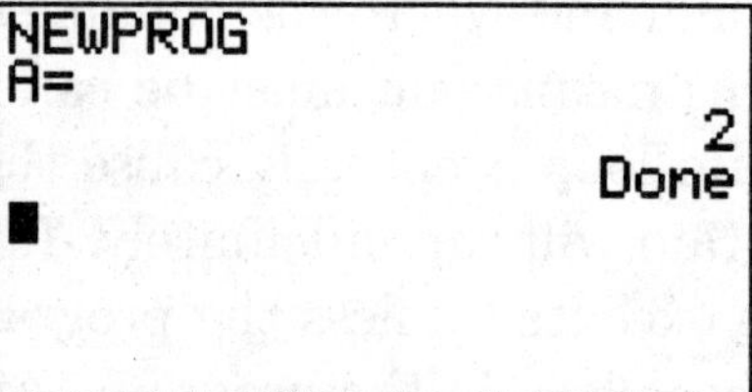

The variable A is started with a value of one, and even though the condition is true, the command to add one to A is executed once before the **Repeat**

loop is terminated. Consequently, the variable A has a final value of 2. In this case, the **Repeat** loop only gets evaluated once.

Another approach to controlling the flow of a program is to use the **Lbl** (for label) ($\boxed{\textbf{MORE}}$ $\boxed{\textbf{F4}}$) and **Goto** ($\boxed{\textbf{MORE}}$ $\boxed{\textbf{F5}}$) commands. The **Lbl** command allows you to give a name to a particular line in your program. The name can be up to eight characters long. The **Goto** command has a label as an argument and will redirect the program to that label. You should be careful in using **Goto** commands as they are always executed. There is no condition to check to see what should be done. The screen shot shows how to correctly label a command and use the **Goto** command.

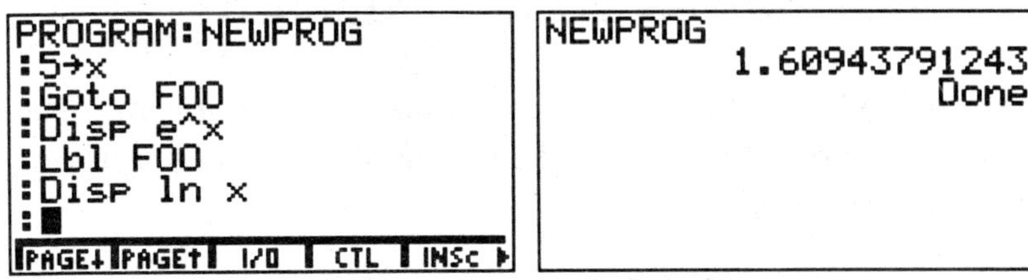

In this simple program, the name FOO is used to label a particular line in the program. When the **Goto** line is reached, the program is immediately redirected to the next command after the label, which in this case displays the value of the natural log of five. Note that the exponential will never be evaluated.

It is good programming style to break up repeated tasks into subroutines. A subroutine is essentially the same as a program except it must end with the **Return** ($\boxed{\textbf{MORE}}$ twice and then $\boxed{\textbf{F4}}$) command. A name is given to the subroutine and it is edited the same as a program. To call a subroutine from a program, the command line is just the name of the subroutine. The program will temporarily leave its list of commands and do those in the subroutine. It will return to the main program when the **Return** command is reached. In the example, the main program **MAIN** calls the subroutine **GOO** which squares x and then returns to the main program where the result is displayed.

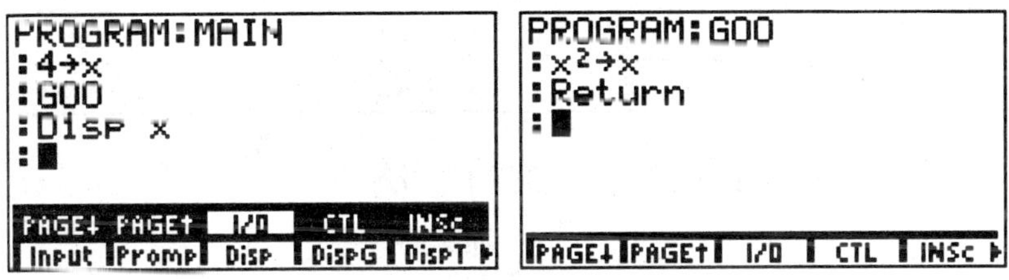

The last commands we discuss are **Pause** (|MORE| twice and then |F3|)
and **Stop** (|MORE| twice and then |F5|). Pause will temporarily stop the
execution so you can see the current screen. You will know that the pause
mode is active due to the dotted busy bar indicator in the top right corner
of the display. Note that is different from the usual busy indicator. **Pause**
commands can be used to troubleshoot a program by placing them at key
points in the program. The program is restarted by pressing |ENTER|. The
Stop command immediately stops the program.

9.5   Infinite Series Plot Program

The program below gives the user a nice tool for studying infinite series. It
allows the user to enter a formula for the sequence of elements to be summed.
The result is plotted on the graphing screen.

:ClLCD	Clears display screen
:Func	Sets graph type to function
:Input ("START=", I)	Prompts user for starting index value
:Input ("END=",J)	Prompts user for ending index value
:InpSt "ENTER SEQUENCE:", Z)	Prompts user for a sequence formula
:St►Eq(Z, y1)	Convert string to equation
:seq(y1, x, I, J, 1)→A	Evaluates terms I to J, stores in A
:A(1)►S(1)	First partial sum stored in list S
:For (K, 2, J−I+1, 1)	For loop to do all partial sums
:A(K)+S(K−1)►S(K)	Find next partial sum
:End	Loop completed
:"A(x)"►AA	Store A(x) as a string
:St►Eq(AA, y1)	Convert string to y1=A(x)
:"S(x)"►BB	Store S(x) as a string
:St►Eq(BB, y2)	Convert string to y2=S(x)
:1►xMin	Set xMin to 1
:127►xMax	Set xMax to 127
:DispG	Display graphs of A(x) and S(x)

This program can be used to study the convergence of the series

$$\sum_{k=2}^{\infty} \frac{1}{k \ln k}.$$

The program first prompts for the starting and ending values of the summation index. For this series, the starting index is 2 and the ending index can be anything larger than or equal to 128. This value of course corresponds to the fact that there are 126 pixels along the width of the display. This choice allows you to view the partial sums nicely on the graphing screen. If you choose a value larger than 128 you will be able to examine the graph of the partial sums further out in the sum. Next you are prompted for a sequence formula. This is just the terms in the series. Since this will become the function **y1**, it must be written as a function of x and not as a function of the index k in this example. Consequently, you should enter $1/(x \ln x)$ for the sequence formula. Now the program has everything it needs to begin.

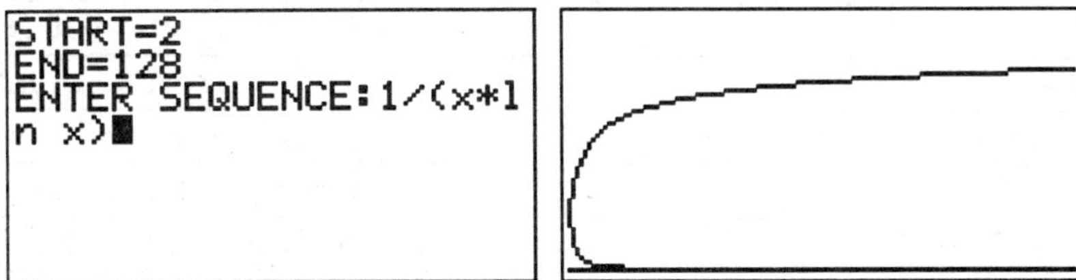

The sequence elements are calculated using the **seq** command and the result is stored as the list A. A **For** loop is used to construct the partial sums which are stored as the list S. This proves be much more efficient than using the **sum** and **seq** commands. The functions corresponding to A and S are set to **y1** and **y2** respectively. Since lists are always indexed starting at one, the values of $xMin$ and $xMax$ are set to 1 and 127, even though the original series starts at 2. Lastly the graphs of the sequence elements and the partial sums are displayed. The user can then use the **TRACE** or **ZOOM** commands to study these graphs in more detail.

9.6 Simpson's Rule Program

Even though, the TI-86 has a built-in integrator for evaluating definite integrals, it is instructive to have an independent program that also does this. With the built-in integrator all the details of how the calculation is done are hidden from the user. With this program, one can examine many of the details.

Simpson's rule is used to evaluate the definite integral

$$\int_a^b f(x)\, dx$$

by first dividing the region of integration into N equal subintervals of width H, where N is an even number. On each subinterval the function's graph is approximated by a parabola. The result is that

$$\int_a^b f(x)\,dx \cong \frac{H}{3}\left(f(x_0) + 4f(x_1) + 2f(x_2) + \cdots + 4f(x_{n-1}) + f(x_n)\right)$$

where $x_0 = a, x_1 = a + H, x_2 = a + 2H, \ldots, x_n = b$. The details of the derivation can be checked in Section 8.7 of Stewar's **Calculus**.

:InpSt "ENTER f(x):", Z	Enter the integrand
:St▶Eq(Z,y1)	Convert string to equation
:Prompt A, B, ERR	Prompt for limts and error
:10▶ N	Set starting value of N
:10E99▶ I	Set starting value of integral
:Repeat abs$(I - J) < ERR$	Test for error condition
:$I \blacktriangleright J$	Set comparison value
:$(B - A)/N \blacktriangleright H$	Evaluate grid width
:$A \blacktriangleright x$	Set first value of x
:For (K, 1, $N + 1$, 1)	For loop
:y1▶Y(K)	Evaluate function
:$x + H \blacktriangleright x$	Next grid point
:End	End of loop
:$Y(1) + 4 * Y(N) + Y(N + 1) \blacktriangleright I$	Start Simpson's rule
:For (K, 1, $N/2 - 1$, 1)	For loop
:$I + 4 * Y(2 * K) + 2 * Y(2 * K + 1) \blacktriangleright I$	Simpson's rule
:End	End of loop
:$2 * N \blacktriangleright N$	Double number of grid points
:$I \star H/3 \blacktriangleright I$	Simpson approximation
:End	End of Repeat loop
:Disp I	Display answer

When this program is run, you are prompted to enter an integrand. This must be a function of x and not some other variable name. You are also prompted to enter the integration limits.

```
SIMPSON
ENTER f(x):x^9
A=?0
B=?1
ERR=?.001
             .100017344242
                     Done
```

The program starts with 10 divisions of the integration region. The value for ERR, which is a measure of how close successive approximations to the integral are, is specified by the user. Each successive iteration doubles the number of divisions. The repeat loop compares successive calculations to see if they are within the prescribed error. You may do significantly better than the value of ERR that was entered, as it can be shown that an error bound for Simpson's rule is inversely proportional to $1/N^4$. So doubling N can be a significant improvement. Lists are used to store the function values at the grid points.

9.7 Exercises

1. Write a program that makes use of the built-in **evalF** command to evaluate functions that are defined piecewise at any point on their domain. To start, assume the piecewise function is of the form

$$f(x) = \begin{cases} g(x) & \text{if } x < a \\ h(x) & \text{if } x \geq a \end{cases}$$

   Your program should allow you to enter $g(x)$, $h(x)$, a and the value of x that f will be evaluated at. Add to your program so that it can handle three function lines in the definition.

2. Write a program that will make use of the built-in **der1** or **nDer** commands to evaluate the derivative of a function that is defined piecewise at any point on their domain. To start, assume the piecewise function is of the form

$$f(x) = \begin{cases} g(x) & \text{if } x < a \\ h(x) & \text{if } x \geq a \end{cases}$$

   Your program should allow you to enter $g(x)$, $h(x)$, a and the value of x that the derivative will calculated at. Be careful about how to handle the point $x = a$. Add to your program so that it can handle three function lines in the definition.

3. Write a program that will make use of the built-in **fnInt** command to evaluate a definite integral of a function that is defined piecewise on any interval. You may assume the pieces of the function are integrable.

To start, assume the piecewise function is of the form

$$f(x) = \begin{cases} g(x) & \text{if } x < a \\ h(x) & \text{if } x \geq a \end{cases}$$

Your program should allow you to enter $g(x)$, $h(x)$ (integrable functions!), a and the endpoints of the integration interval. Add to your program so that it can handle three function lines in the definition.

4. Modify the Simpson's rule program so that the user can control the number of grid points, N and view the approximation due to this approximation for each value of N used.

5. Modify the Simpson's rule program so that it approximates the integral using the Trapezoidal rule given by

$$\int_a^b f(x)\, dx \cong \frac{H}{2}\left(f(x_0) + 2f(x_1) + 2f(x_2) + \cdots + 2f(x_{n-1}) + f(x_n)\right)$$

Chapter 10

Labs

10.1   The Ant and the Blade of Grass

Prerequisites: Read the introductory material in chapters 1 and 2, and read about tangent lines in chapter 3 of this manual. In particular, make sure you can use the TI-86 to graph a function, draw tangent lines and use the **SOLVER** to solve equations numerically. In addition, read Section 2.6 in Stewart's **Calculus**.

Background: An ant is walking (to the right) over its ant mound, whose height (in inches) is given by the function

$$h(x) = \frac{x^2/16 - 2x + 80}{(x^2/16 - 2x + 20)^2}.$$

Nearby there is a blade of grass, which is located as the line segment from $(32, 1/5)$ to $(32, 8)$.

Assignment: Find the point where the ant first sees the blade of grass. You can assume that the ant's line of sight is the tangent line to the ant mound. The following set of questions will lead you to the answer.

1. Define the function $h(x)$ in the graphing screen. Choose a window so that you can view the whole ant mound. Use the **DRAW LINE** command to draw the vertical line that represents the blade of grass.

2. Draw the tangent line to $h(x)$ at $x = 12.5$ using the **TanLn** command. Can the ant see the blade of grass from this point?

3. Find the height where the tangent line crosses the line $x = 32$. You may want to define another function as the tangent line at 12.5 and use **TRACE** to find the intersection.

4. Draw the tangent line to $h(x)$ at $x = 15.5$ using the **TanLn** command. Can the ant see the blade of grass from this point? Find the height where the tangent line crosses the line $x = 32$.

5. From the information found in the previous questions, it is the case that the ant can just see the top of the blade of grass at $x = a$ where $12.5 < a < 15.5$. Write down the equation of a tangent line that passes through the point $(a, h(a))$ and $(32, 8)$.

6. Use the **SOLVER** to solve for the value of a where the equation is the one you wrote down for the tangent line in part 4.

7. Use the value of a found to draw the tangent line on the same display as the blade of grass and the ant mound. Find the height where the tangent line crosses the line $x = 32$.

8. There is a second solution for a that can be found using the **SOLVER**. Find its value. What is wrong with this solution?

10.2   A Power Relay Station

Prerequisites: Setting up word problems and graphing functions. Read Chapter 1 in Stewart's **Calculus**.

Assignment: There are 15 communities located along a straight highway at mile markers 1, 1.5, 2, 2.35, 5, 5.1, 6.35, 6.45, 6.5, 7.33, 9.6, 10.11, 12.5, 14.7, 16. The electric company which serves these 15 communities would like to construct a power relay station somewhere along this highway. The company would then run separate power lines from the relay station to each of the 15 communities. To minimize the cost of constructing these 15 power lines, the company would like to locate the relay station so as to minimize the sum of the distances to each station.

Assignment. Answer the following questions.

1. Where should the electric company build its relay station? Justify your answer. Is this location unique?

2. Suppose a 16th community, located at mile 20, is added to the highway. Now where should the electric company locate its relay station? Is this location unique?

Hint: Think of the communities as located along the x-axis at the mile markers that are given. Suppose the relay station is located at position x. Write the sum of the distances from the power station to the communities in terms of x.

10.3   Tangent and Normal Lines

Prerequisites: Computing tangent lines and normal lines to curves. The second question posed in this project is considerably more difficult than the first. Read Sections 2.6, 3.1 and 3.3 in Stewart's **Calculus**.

Assignment: Answer the following two questions.

1. Show that every point (x, y) with $y < x^2$ has the property that it belongs to two distinct tangent lines to the curve $y = x^2$.

2. Find the equation(s) of the curve C with the property that each point on C belongs to two distinct normal lines to $y = x^2$.

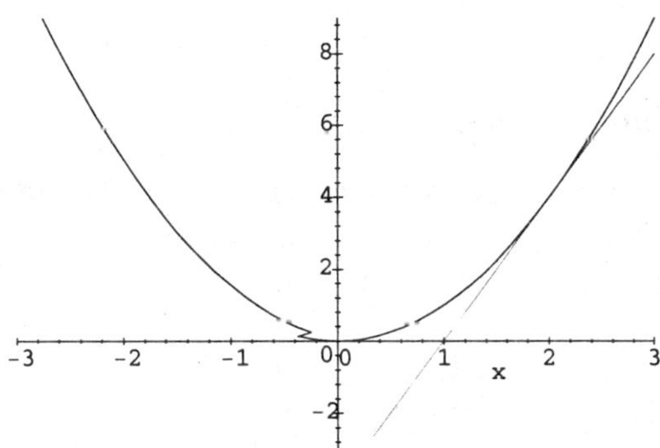

Hints: For the first question, if (a, b) belongs to the tangent line to $y = x^2$ at (x, x^2), then x is the solution of a quadratic equation obtained by setting the slope of the chord from (x, x^2) to (a, b) equal to the derivative of $y = x^2$

(draw a picture). The analogous equation for the second question, involving normal lines, is of the form $f(x) = 0$ where $f(x)$ is a cubic polynomial. In order for a cubic equation to have exactly two roots, one of them, say $x = c$, must also be a root of the derivative of f. Therefore in order for (a, b) to belong to a normal line to $y = x^2$, the resulting equations, $f(x) = 0$ and $f'(x) = 0$, must both have a root at the same point (namely, $x = c$).

10.4   Epicycloid Curves

Prerequisites: Read about parametric equations in Chapter 2 of this manual. Make sure you can use the TI-86 to graph curves parametrically, draw tangent lines and use the **SOLVER** to solve equations numerically.

Background: If you take two circles, one with radius a and the other radius b (with $a \geq b$), and mark the smaller one at a point on the circumference, then a curve is traced out by this point as the small circle rolls about the larger one without slipping. The resulting curve is called an epicycloid. This curve is given parametrically by the equations

$$x = (a + b) \cos t - b \cos \left(\frac{a + b}{b} \right) t$$

$$y = (a + b) \sin t - b \sin \left(\frac{a + b}{b} \right) t$$

where $-\pi \leq t \leq \pi$. These curves can be easily plotted along with the fixed circle parameterized by $x = a \cos t$, $y = a \sin t$ for $0 \leq t \leq 2\pi$.

Assignment: Complete the following exercises.

1. Choose a suitable viewing window and plot the epicycloid when $a = b$. Do the graphs change for different value of a? Is the curve symmetric about any of the axes?

2. Now take a and b to be integers and plot several examples of these curves. What properties do the curves have and is their any correlation between these properties and the values of a and b? Is the curve symmetric about any of the axes?

3. Answer the same type of questions when a and b are rational numbers. Take for example $a = 1$ and $b = 1/2, 1/3, 2/3, 4/5$. Compare the cases when a is an integral multiple of b and when it is not.

4. What happens to the curve when a and b are irrational numbers? As an example take $a = 1$ and $b = \sqrt{2}$.

10.5 Constrained Extremum

Prerequisites: Read about max/min problems in Chapter 4 of this manual. In addition, see Sections 4.1 and 4.7 in Stewart's **Calculus**.

Background: The typical optimization problem has only a single constraint. In this project, you will examine what happens when there is more than one constraint. This problem can be solved using a mostly graphical approach, along with standard calculus techniques for locating extrema.

Assignment: A container in the shape of a right circular cylinder is to be made from material that costs \$1.00 per square inch for the top and bottom, and \$0.50 per square inch for the lateral surface.

1. Find the dimensions of the cylinder that maximizes the volume assuming the cost for the materials to make the cylinder fixed at \$300.

2. Now consider the reverse problem, where the cost is to be minimized with the volume fixed at 750 cubic inches.

3. Find the dimensions of the cylinder that maximizes the volume with the cost for materials at most \$500.

4. Find the dimensions of the cylinder that maximizes the volume with the cost for materials at most \$500, given that the container must fit in an 8 inch wide space.

10.6 Introduction to Hyperbolic Functions

Prerequisites: Exponential functions, limits and differentiation. Read the introductory material in chapters 1, 2 and 3 of this manual. In addition, read Chapters 1, 2, 3 and 7 of Stewart's **Calculus**.

Background: The hyperbolic functions are defined in terms of exponential functions, but they are of interest in their own right. The three basic hyperbolic functions are given by

$$\sinh x = \frac{e^x - e^{-x}}{2}$$

$$\cosh x = \frac{e^x + e^{-x}}{2}$$

$$\tanh x = \frac{e^x - e^{-x}}{e^x + e^{-x}}$$

The names are *hyperbolic sine, hyperbolic cosine* and *hyperbolic tangent*. The connection with the standard trigonometric functions can be seen when these functions are studied in the complex plane. One important application is the shape that a cable or chain hangs between two supports.

Assignment:

1. Graph the three hyperbolic functions. What is the domain and range for each?

2. Graph $\sinh$, $\cosh$ and their derivatives by using the **der1** command. Make a conjecture about the derivatives and then use the definitions of the hyperbolic functions to prove the conjecture is true. Also plot the quantity $\cosh^2 x - \sinh^2 x$. Make a conjecture about this as well, and then use the definitions of the hyperbolic functions to prove your conjecture..

3. We say two functions, $f(x)$ and $g(x)$, are asymptotic to each other in the vicinity of a point $x = a$, written $f(x) \sim g(x)$ as $x \to a$, if

$$\lim_{x \to a} \frac{f(x)}{g(x)} = 1.$$

   This means the two functions are roughly the same in value near $x = a$. By graphing in an appropriate window illustrate the fact that $\sinh x \sim e^x/2$ and $\cosh x \sim e^x/2$ as $x \to \infty$. Also verify this by evaluating the limits.

4. When a cable is suspended between two poles, it's shape is given by a hyperbolic cosine function. Suppose that the equation of the curve for the shape of the cable is given by $y = a\cosh(x/a)$. Given that the poles are located at $x = \pm b$ show that the length of the cable is

$2a\sinh(b/a)$ where a and b are both positive constants. The sag of the cable is defined by the vertical distance between the highest and lowest points on the cable. Show that the sag is given by $S = a\cosh(b/a) - a$.

5. Suppose the distance between power line poles is 100 meters and the allowable sag is 3 meters. Find the value of a that determines the equation of the power line and the length of cable required.

10.7   The Car Trip

Prerequisites: Velocity, speed, acceleration and total distance travelled. See Chapters 4, 5 and 6 in Stewart's **Calculus**.

Background: On a nice warm summer day, Tom and Sandi decide to go for a drive in the country. They take the new highway to avoid the many traffic lights along the other route. They leave Hipsville at twelve noon and continue driving through to the town of Squaresville. Since there is nothing to do there, they immediately turn their car around and head back. Before reaching Hipsville, they both get hungry and stop at a restaurant at 7 PM for a snack. Sandi, being a physics major, had decided to take speedometer readings every few minutes along the whole trip. While waiting for their food at the restaurant, they decide to fit a polynomial to the data. They find the velocity in miles per hour of the car as a function of the time in hours was given by

$$v(t) = -\frac{1}{3}t^5 + 6t^4 - 35t^3 + \frac{196}{3}t^2$$

where positive values of v correspond to the outgoing trip and negative values to the return trip.

Assignment: Answer the following questions.

1. At what time do they reach Squaresville?

2. The speed is given by $s(t) = |v(t)|$. Plot a graph of the speed. Find the maximum speed for each part of the trip.

3. The distance travelled by the couple is given by

$$d(t) = \int_0^t |v(\tau)|\, d\tau.$$

How far is it from Hipsville to Squaresville? How many miles away were Tom and Sandi from home when they stopped at the restaurant?

4. The average of a function $f(t)$ on an interval $[a, b]$ is given by

$$f_{AVG} = \frac{1}{b-a} \int_a^b f(t)\, dt.$$

Find the average speed on the drive out to Squaresville. Find the average speed on the trip till the restaurant stop.

5. The acceleration $a(t)$ of the car is given by the time derivative of the velocity. The car is accelerating if $a(t) > 0$ and decelerating if $a(t) < 0$. Find an equation for the acceleration as a function of time. Find the time intervals over which the car was accelerating and decelerating.

10.8   Normal Distributions

Prerequisites: Read about graphing in Chapter 2 and integration in Chapter 5 of this manual.

Background: Shoe sizes, height and test scores are some examples of phenomena that are typically modeled by a normal distribution, or what is often called a bell curve. The normal distribution is given by the function

$$f(x) = \frac{1}{\sigma\sqrt{2\pi}} e^{-(x-\mu)^2/2\sigma^2}$$

where σ and μ are two parameters that are called the standard deviation and the mean respectively. The probability that a random variable X lies between two numbers is given by

$$P(\alpha \le X \le \beta) = \int_\alpha^\beta f(x)\, dx.$$

Assignment: Complete the following exercises.

1. Plot the normal distribution for $\sigma = 1$ and several values of μ by defining a function like

$$y1 = e\,\hat{}\,((x - \{1, 2, 3\})^2/2)$$

where the $\{1, 2, 3\}$ denotes a graph list of values for μ. What is the graphical interpretation of the mean?

2. Plot the normal distribution for $\mu = 0$ and several values of σ. What is the graphical interpretation of the standard deviation?

3. Show that the area under this curve is independent of the mean and the standard deviation by using the change of variable

$$z = \frac{x - \mu}{\sqrt{2}\sigma}$$

Use the fact that

$$\int_{-\infty}^{\infty} e^{-x^2}\, dx = \sqrt{\pi}$$

to evaluate the integral

$$\int_{-\infty}^{\infty} \frac{1}{\sigma\sqrt{2\pi}} e^{-(x-\mu)^2/2\sigma^2}\, dx$$

4. Suppose the average test score on a standardized test is 125 out of 200 and the standard deviation is 20. What percentage of the people taking the test scored between 85 and 145.

5. For the same test data, find the range of scores about the mean that 50% of the people obtained. That is, find d so that

$$P(125 - d \le X \le 125 + d) = 0.5.$$

You will need to use the **SOLVER** and the **fnInt** commands to do this.

10.9   Perimeter of an Ellipse

Prerequisites: Parameterized curves and arclength. Read Chapter 11 in Stewart's **Calculus**.

Background: An ellipse is described parametrically by

$$x(t) = a\cos t \quad y(t) = b\sin t, \quad \text{for } 0 \le t \le 2\pi$$

where a and b are the lengths of the major and minor axes, respectively. A circle is a special case of an ellipse when $a = b$.

The perimeter of circle is given by the very well known formula, $P = 2\pi r$ where r is the radius of the circle. For an arbitrary ellipse there is no simple formula like there is for a circle. The orbits of the planets and some of the comets are elliptic in shape.

Assignment: Complete the following exercises.

1. Use the parametric plot routines to plot several ellipses for different choices of a and b.

2. Show that the equation for the perimeter of an ellipse is

$$P = 4 \int_{-\pi/2}^{0} \sqrt{a^2 \sin^2 t + b^2 \cos^2 t}\, dt$$

$$= 4a \int_{0}^{\pi/2} \sqrt{1 - \epsilon^2 \sin^2 \tau}\, d\tau$$

where the change of variable $\tau = t + \frac{\pi}{2}$ has been used. The parameter ϵ is called the eccentricity and is defined by

$$\epsilon = \frac{\sqrt{a^2 - b^2}}{a}$$

The eccentricity is a measure of the shape of the ellipse. If $\epsilon = 0$, then the ellipse is a circle. When a is much larger than b, ϵ is approximately 1, and the ellipse is very narrow. Verify that the formula for the perimeter of an ellipse reduces to the correct formula when $\epsilon = 0$ which is a circle.

3. The integral in the formula for P is called an elliptic integral and must be evaluated numerically as an anti-derivative cannot be found in terms of elementary functions. Evaluate this integral for several choices of a and b.

4. The Earth's orbit is almost circular with $a = 93$ million miles and $\epsilon = 0.017$. Find the distance travelled by the Earth during one year. Pluto's orbit is more eccentric with $\epsilon = 0.249$ and $a = 4900$ million miles. Find the distance travelled by Pluto during one circuit about the sun.

5. Define an integral defined function with ϵ as the variable and graph

$$E(\epsilon) = \int_0^{\pi/2} \sqrt{1 - \epsilon^2 \sin^2 \tau}\, dt$$

on the interval $0 \le \epsilon < 1$. This plot will allow you to find the perimeter of any ellipse. What is the perimeter when $\epsilon = 0.5$.

6. A planet that has an elliptical orbit with time period T (in years) can by represented parametrically by

$$x(t) = a\cos(2\pi t/T) \qquad y(t) = b\sin(2\pi t/T).$$

Assume that the Earth's orbit is circular, with its orbit given by

$$x(t) = \cos(2\pi t) \qquad y(t) = \sin(2\pi t)$$

where the units of length are the Earth-Sun distance, which is referred to as one astronomical unit (AU), and t is measured in years. Suppose a comet with a period of 10 years follows the elliptic path given by

$$x(t) = 9.5 + 10\cos(\pi t/5) \qquad y(t) = 0.5\sin(\pi t/5)$$

with $0 \le t \le 10$. During one period of the comet find the nearest and farthest distances in AU between the two bodies. Does the comet collide with the Earth?

10.10 Soap Bubbles

Prerequisites: In this lab, you will use the TI-86 to evaluate surface area and examine a famous problem from the calculus of variations. Some knowledge of hyperbolic functions is required. Read about the **SOLVER** in chapter 1 and integration in chapter 5 of this manual. In addition, see Chapters 5 and 7 in Stewart's **Calculus.**

Background: As a child you probably used a wand with a circular ring for blowing soap bubbles. In this project we examine the shape of the bubble formed when two wands with circular rings of possibly different radii, are dipped together in a soap solution and then are slowly pulled apart. The result is a cylindrical type bubble that connects the two wands. Of course, if you pull the wands too far apart the bubble will break. Physically the shape of the bubble is determined by the requirement that the surface area is minimal, which is effectively a minimum energy configuration. Mathematically, we can think of the bubble shape as defined by revolving the graph of a function $f(x)$ about the x-axis to form the volume of revolution. The surface of this volume is the soap bubble. The requirements on the function are that the graph passes through the points $(0, R_1)$ and (L, R_2), where R_1 is the radius of the wand on the left, R_2 is the radius of the wand on the right, and L is the distance separating the wands. The surface area of the volume of revolutions can be shown to be

$$S = 2\pi \int_0^L f(x)\sqrt{1 + f'(x)^2}\,dx$$

The soap bubble shape chooses a function $f(x)$ which minimizes this integral. The function $f(x)$ that minimizes this integral can be found using techniques from the calculus of variations. In this project, you will calculate the surface area for several soap bubbles and see that the actual shape that minimizes the surface area is given by the surface known as a catenoid.

Assignment: Complete the following exercises.

1. Assume that the radii of the rings on the wands are $R_1 = 1cm$ and $R_2 = 2cm$, and the length between the wands is $L = 1cm$. Sketch a picture of the bubble problem labeling all the relevant points. Find a simple integral formula for the surface area for the following shape bubbles: a) $f(x)$ is a straight line connecting the endpoints, b) $f(x)$ is a parabola with vertex at the left-hand endpoint, and c) $f(x)$ is a quarter circle with center $(1, 1)$ and radius 1. Plot graphs of all three on the same axes. Predict which will lead to the smallest and largest surface area.

2. Evaluate the three integrals in the exercise above. Perform at least one of the integrals analytically. Which of the shapes gives the smallest surface area? The largest surface area?

3. Using the calculus of variations, it can be shown that the function that minimizes the surface area is a hyperbolic cosine function in the form

$$f(x) = \alpha \cosh\left(\frac{x + \beta}{\alpha}\right)$$

where α and β are constants. Plot this function for several choices of the constants. The values of α and β should be chosen so that the graph passes through the two endpoints. The surface of revolution for a hyperbolic cosine is called a catenoid.

4. Show that the surface area is given by the simple integral formula

$$S = 2\pi \int_0^1 \cosh^2\left(\frac{x+\beta}{\alpha}\right) dx$$

5. The endpoint conditions $f(0) = 1$ and $f(1) = 2$ give the equations

$$1 = \alpha \cosh\left(\frac{\beta}{\alpha}\right)$$
$$2 = \alpha \cosh\left(\frac{1+\beta}{\alpha}\right)$$

for the α and β. Use these two equations to derive the single equation for α:

$$2 = \alpha \cosh\left(\frac{1 + \alpha \cosh^{-1}(1/\alpha)}{\alpha}\right)$$

This equation must be solved numerically. Use this equation to find the numerical values of α and β. Plot the resulting function and verify that it passes through the correct endpoints. With these values find the surface area of the bubble. How does it compare with the other shapes considered?

Further Exploration: Explore what happens to the catenoid as the length separating the endpoints increases. See if the bubble breaks (mathematically) after a critical value of the length.

10.11 The Brightest Phase of Venus

Calculus Prerequisites: This is the hardest project - nontrivial trigonometry, max/min theory, area. Read Sections 4.7, 5.1, 5.2 and 6.1 in Stewart's **Calculus**.

Background: The brightness of Venus is proportional to the area of the visible portion of Venus and inversely proportional to the square of the distance from the Earth to Venus. From the accompanying figure, note that, as the angle t increases from 0 to π, the area of the visible portion of Venus increases. This tends to increase the brightness of Venus. But the distance d from the Earth to Venus also increases, which tends to decrease the brightness of Venus. For some angle t, between 0 and π, Venus will appear brightest.

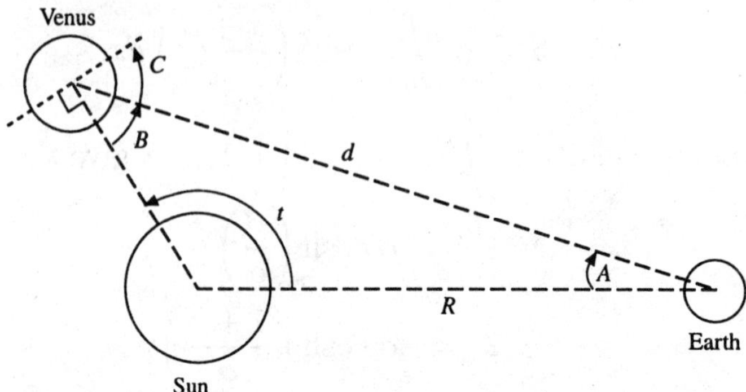

Assignment: Find the brightest phase of Venus (i.e., the angle t) by following the outline given below. Write up your solution using complete sentences. Use the TI-86, where appropriate, to help with the computations and graphics.

1. As mentioned above, brightness is proportional to the quantity

$$\frac{\text{Area of visible portion of Venus}}{d^2}$$

2. From geometry, show that $B = \pi - (t + A)$ and $C = (t + A) - \pi/2$.

3. Let a be the radius of Venus. Show that the visible portion of Venus (from Earth) lies between the curves

$$x = -\sqrt{a^2 - y^2} \text{ and } x = \sin(C)\sqrt{a^2 - y^2}$$

Now show that the area of the visible portion of Venus is

$$\frac{\pi a^2}{2}(1 + \sin(C))$$

4. Combine parts 1, 2, and 3 to obtain a formula for the brightness of Venus that depends on the angles t and A and the distance d. The goal is to express the brightness in terms of one variable t. To this end, use the law of cosines and the law of sines to show the following

$$d^2 = r^2 + R^2 - 2rR\cos(t)$$

$$\sin(A) = \frac{r\sin(t)}{d}$$

   Here, r is the distance from the Sun to Venus and R is the distance from the Sun to the Earth. Use the values $R = 93$, $r = 67$, and $a = 0.004$ (the unit of distance is one million miles).

5. Use the equations in part 4 to find an expression for the brightness of Venus that depends on the single variable t. Use the TI-86 to maximize this function over the interval $0 \le t \le \pi$.

10.12 Evaluating Definite Integrals in Monte Carlo

Prerequisites: Definite integrals and basic programming. See chapters 5 and 9 in this manual, and Chapter 5 in Stewart's **Calculus**.

Background: One of the things Monte Carlo is famous for is the gambling casino. A technique known as the Monte Carlo method uses simple probability ideas which can be implemented to approximately evaluate a definite integral. Consider a function $f(x)$ with $0 \le f(x) \le 1$ on the interval $[0, 1]$; i.e. the function's graph fits exactly on the unit square. Suppose the graph was been drawn on a piece of graph paper with the unit square having a 50 by 50 grid marked on it. Now take two standard roulette wheels (from the Monte Carlo casino of course) which have numbers zero through 48 plus a double zero, and play them both. Assuming these are fair games of chance, the two numbers that come up when the marble stops rolling on each wheel give a unique position on this grid. Since any number is equally likely to occur, any of the 2500 grid locations can be selected. Mark the spot on the graph paper and continue this process many times. Next count how many marks are under of the graph of $f(x)$. Since locations on the square are equally likely to be chosen, the percentage of those under the graph should be the same as the percentage of the unit square that is the area under the curve. Consequently, the ratio of this number to the total number of marks is an estimate of the area under the graph. This area is also given by the integral

$$\int_0^1 f(x)\,dx$$

Of course, to get reasonable answers, you must play this game many times. Unfortunately, you would surely be bored before you got a reasonable answer. Fortunately, a computer or calculator can be used to generate random numbers, and it won't get bored. The program given below generates random coordinates on the unit square and keeps track of how many are under the graph of a given function. The command **rand** can be accessed by pressing [2nd] [MATH] and then [F2] for **PROB**. This gives the probability set of commands. Press [F4] to generate a random number between zero and one.

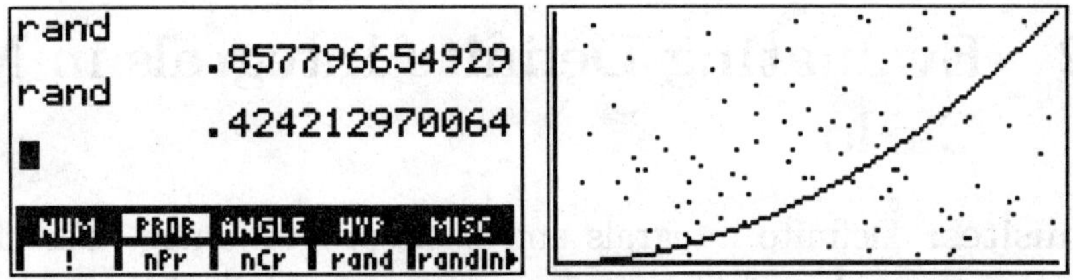

The screen shot shows the result of using the program to generate 100 random coordinates on the unit square for the function $f(x) = x^2$. Use the program below to approximately evaluate integrals which can be compared with exact and numerical results.

Assignment:

1. Carefully copy the program **MCARLO** (given at the end of this assignment) into your calculator's memory.

2. Before running the program be sure to set the window so that it is the unit square. The screen shot shows how the program runs. You will be prompted to enter a function. It's graph should exactly fit inside the unit square. You will also be prompted to enter a value for N, which is the number of random points to generate. To start, choose a small value to make sure the program is working correctly. Be aware that if

you choose N to be very large, you may wait a long time to get the result. If you want to interrupt the program, press $\boxed{\text{ON}}$ while it is running and you will have the option of quitting the calculation.

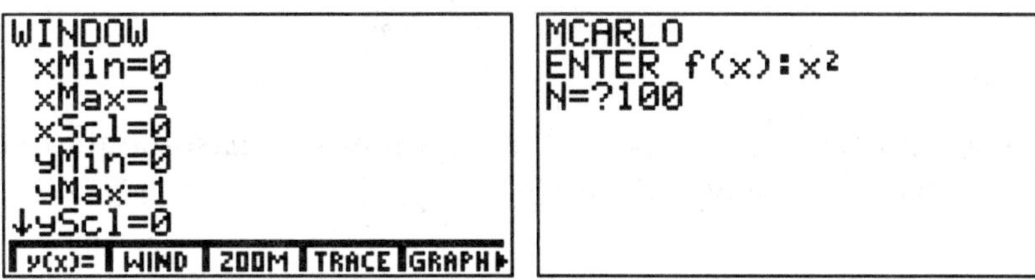

The program will first draw the graph of the function and then start plotting the randomly generated points. When they are all plotted, the approximation to the integral is displayed. If you want to see the graph again, press $\boxed{\textbf{GRAPH}}$.

3. Use $f(x) = x^2$ to check to make sure the program is working correctly. Choose values of N of 10, 20, 50 and 100. You should get reasonably good approximations to the exact value

$$\int_0^1 x^2 \, dx = \frac{1}{3}$$

for the larger values of N.

4. Take $f(x) = \sin(\pi x/2)$ and find approximate values of its definite integral on the interval $[0, 1]$. **Make sure your calculator is in radian mode!!** Take N to be 100 and run the program several times. Do you always get the same approximation to the exact value? If not, explain why.

5. Approximate the definite integral

$$\int_0^{\pi/2} \left(1 - \frac{1}{4}\sin^2\theta\right)^{1/2} d\theta$$

by the Monte Carlo method. You will first have to rescale the integration variable so that the integration interval is $[0, 1]$ before using the program. This integral cannot be evaluated exactly with elementary functions, so you should compare your results with that given using numerical integration with the **fnInt** command. How do the results compare?

6. Think of how you could use this program to find approximate values of any integral of the form

$$\int_a^b f(x)\, dx$$

You may assume $f(x) \geq 0$ but you may not assume its maximum value is 1. Try your method on the integral

$$\int_{-1}^3 (3 + 2x - x^2)\, dx$$

The program **MCARLO**:

InpSt "ENTER f(x):", Z	Input function $f(x)$
St►Eq(Z, y1)	Store function as y1
Prompt N	Input number of total points
0►J	Reset J to be zero
ClDrw	Clear graphing screen
For(I, 1, N, 1)	Set loop to generate numbers N times
rand►x	Generate random number for x
rand►Y	Generate random number for Y
PtOn(x,Y)	Plot the coordinates on graph
If Y≤y1	Compare Y coordinate with function value
J+1►J	If Y≤y1 then increment J
End	End of loop
Disp J/N	Display the result

10.13   Center of Gravity

Prerequisites: Computing normal lines to a curve and center of mass. Read Section 2.6 and Chapter 5 in **Calculus: Concepts and Contexts, Single Variable**.

Background: The center of mass of a region bounded above by $y = f(x)$ and below by $y = g(x)$, $a \leq x \leq b$, is the point $(x0, y0)$ where

$$x0 = \frac{1}{A} \int_a^b x[f(x) - g(x)]\, dx$$

$$y0 = \frac{1}{A} \int_a^b 1/2 \left([f(x)]^2 - [g(x)]^2\right) \, dx$$

Consider a thin metal plate which is cut in the shape of a parabola bounded by the parabola $y = \frac{1}{40}x^2$ and the line $y = 10$. Suppose the x-axis corresponds to ground level and the y-axis corresponds to up and down. If the density of the plate is uniform, then the center of gravity will be located on the y-axis (at $(0,6)$). If the plate is rolled along the x-axis slightly, the plate will become "off-balance" and will try to roll to a new equilibrium state.

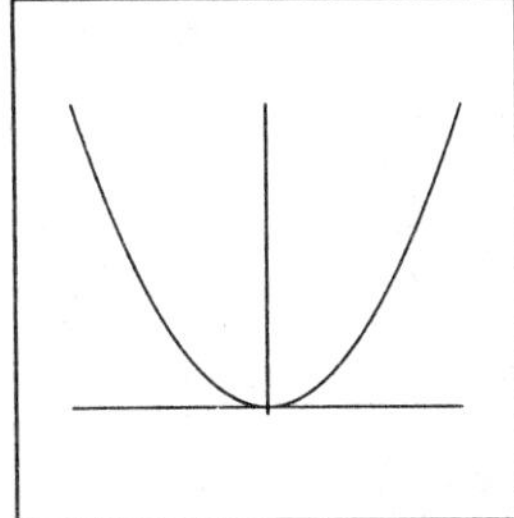 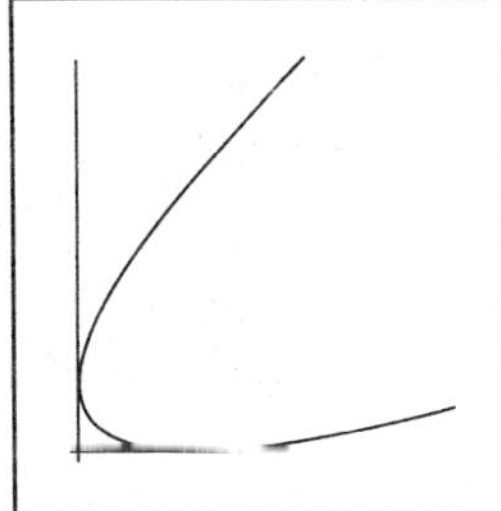

Assignment: Consider the thin metal plate described above. Suppose the plate is pushed very slightly so that it rolls in the positive x direction.

1. Give a geometric argument which explains why the plate will not return to its original position.

2. Find the new equilibrium state that the plate will move towards.

3. Give the coordinates of the center of gravity of the plate in its new equilibrium position.

4. Repeat problems 2 and 3 for the generic parabola $y = ax^2$ where $a > 0$ (still use the line $y = 10$).

5. Give a parametric plot (in terms of the parameter a) of the center of gravity found in problem 4.

10.14 Curves Generated by Rolling Circles

Prerequisites: Parameterized curves, the cycloid and arc length. Read Chapter 11 in Stewart's **Calculus**.

Background: The goal of this project is to use the TI-86 plotting and integrating capabilities to help answer questions about the cycloid; a curve generated by a point on a rolling circle. A more complicated version of the cycloid is also considered.

Consider a wheel of radius R. Fix a point on the rim of the wheel. Now let the wheel roll on level ground and consider the path traced out by the point P (see the figure). This path is called the cycloid.

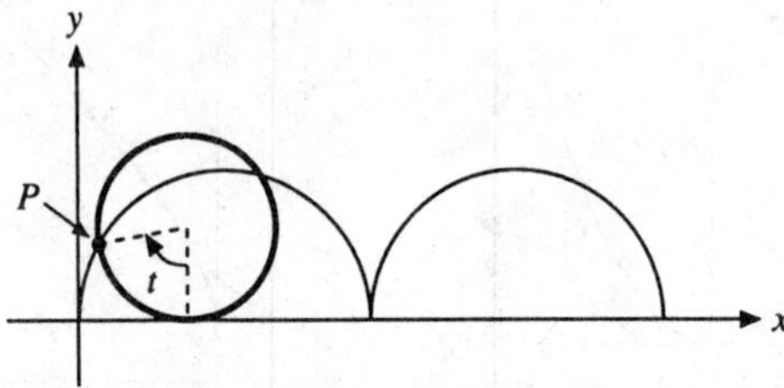

Assignment: Answer the following questions, using the TI-86 where appropriate.

1. Show that the cycloid is parameterized by the formulas

$$x(t) = R(t - \sin(t)) \text{ and } y(t) = R(1 - \cos(t))$$

   Here, t is the angle between the vertical and the ray that extends from the center of the circle to P (so $t = 0$ when P is at the origin).

2. Use DERIVE to plot two arches of this cycloid with $R = 1$.

3. Compute the arc length of one arch of this cycloid (for a general value of R). Recall that the arc length of a parameterized curve $(x(t), y(t))$ for $a \leq t \leq b$ is given by

$$\int_a^b \sqrt{(x'(t))^2 + (y'(t))^2}\, dt$$

4. In the previous problem, you determined the arc length of one arch by computing the integral over the interval $0 \leq t \leq 2\pi$. Presumably, computing the appropriate integral over the interval $0 \leq t \leq 4\pi$ should give the arc length of two arches. Try this.

5. The slope of a parameterized curve $(x(t), y(t))$ is given by

$$\frac{dy}{dx} = \frac{dy/dt}{dx/dt}$$

Show that the limit of the slope of the tangent line as $t \to 0^+$ is infinity.

Note: It may be easier to show that $1/\text{slope} \to 0$ as $t \to 0^+$.

6. Now suppose a circle of radius a rolls around the outside of the circle of radius $R > a$ centered at the origin (see the accompanying diagram). Find the parameterization $P = (x(t), y(t))$ that describes the path of a fixed point P on the rolling circle. Here t is the angle measured counterclockwise from the positive x-axis to the line segment that runs from the origin to the center of the rolling circle. Assume that P is located at the point $(R, 0)$ when $t - 0$. Compute the arc length of one of the arches of this path.

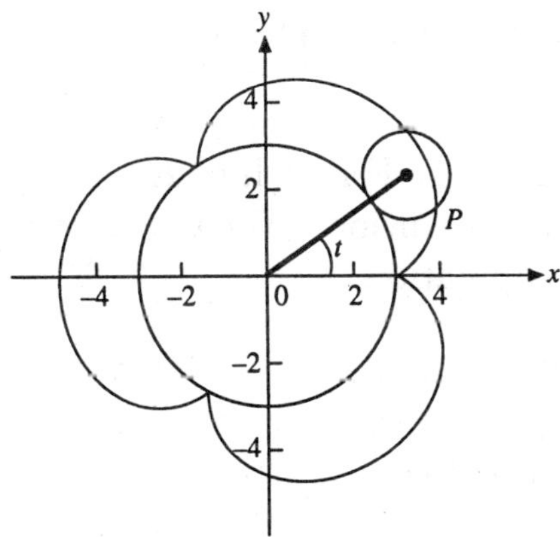

Related Activity: See the Laboratory Project: *Families of Hypocycloids* at the end of Section 11.1 in Stewart's **Calculus**.

10.15   Harvesting a Population of Fish

Prerequisites: Read about differential equations in Chapter 7 of this manual and Chapter 10 of Stewart's **Calculus**.

Background: A population of a particular isolated species is often modeled by the logistic differential equation

$$P'(t) = r(1 - P(t)/K)P(t)$$

where r is the growth rate of the population and K is the carrying population (i.e. the maximum population that can be sustained). Let us suppose that this population refers to fish raised at a fishery, and the model is modified to include harvesting which take place at a constant rate H. Then the new differential equation becomes

$$P'(t) = r(1 - P(t)/K)P(t) - H$$

Assignment:

1. Take $r = 1$, $K = 1$ and $H = 0$, and investigate how the initial condition $P(0) = P_0$ effects the solution. What is the limiting behavior of $P(t)$ for large t?

2. Keep $H = 0$ and $K = 1$. For each of $r = 0.1, 1$ and 2, plot the solution for a fixed initial condition. What can you conclude about how r affects the fish population?

3. Keep $H = 0$ and $r = 1$. For each of $K = 1, 5$ and 10, plot the solution for a fixed initial condition. What can you conclude about how K affects the fish population?

4. Fix $r = 1$ and $K = 5$ and vary the harvesting rate. For what values of H does the fish population survive? Find the critical value of H where the population just becomes extinct.

5. Verify you answers in part 4 analytically by finding the equilibrium solutions to the logistic equation with harvesting. The equilibrium solutions are the constant solutions of the equation which have $P'(t) = 0$.

6. Now consider part 4 in the case of general values of r and K by considering the shape and algebraic sign of the graph of

$$F(P) = (1 - P/K)P - H$$

10.16 Graphing Implicit Functions

Prerequisites: Implicit functions, parametric equations and differential equations. Read chapters 2 and 7 in this manual, and Chapters 10 and 11 in Stewart's **Calculus**.

Background: The graph of an implicit function is not easy to sketch, due to the difficulty of solving complicated equations. One can make a table of x values and then solve the resulting equation for y, either exactly or numerically, but this is a tedious process. It is easier to express the problem as set of differential equations and solve them numerically.

Assignment:

1. Suppose a curve is given by the equation

$$x^3 + y^3 = 3xy + 75$$

   Verify that $(-5, 5)$ satisfies this equation. Assuming this equation defines an implicit function, use implicit differentiation to find dy/dx for this function.

2. If the curve that satisfies this equation is given parametrically by $x = x(t)$ and $y = y(t)$, then

$$\frac{dy/dt}{dx/dt} = \frac{dy}{dx} = \frac{y - x^2}{y^2 - x}.$$

   Equating the numerator and denominator we find two differential equations

$$\frac{dx}{dt} = y^2 - x$$
$$\frac{dy}{dt} = y - x^2$$

   for the functions $x(t)$ and $y(t)$. On what curve in the x-y plane are tangent lines to this curve horizontal? Vertical?

3. Translate these two differential equations into a form handled by the TI-86 calculator by using $Q1$ and $Q2$ for the variables. Set the axes to be $Q1$ and $Q2$. Set the window parameters so that $-5 < x < 5$, $-5 < y < 5$ and $0 < t < 1$. Next choose initial conditions. Since the parametrization of a curve is not unique we will choose:

$$x(0) = -5 \qquad y(0) = 5.$$

Solve the resulting differential equations. Does the resulting curve correspond to a function? If not, explain why? Choose three parts of the curve give a function, and state their domain.

4. Find the equations of each tangent line when $x = 4$, using either the **SOLVER** or **TRACE**.

5. Consider the equation

$$x^3 + y^3 = 3xy + C$$

where C is any real number. Give a differential equation which is satisfied by this implicit equation regardless of the value of C. Use the differential equation solver to plot solutions for different initial conditions. Give the value of C for which these solutions satisfy the equation above.

10.17   The Flight of a Baseball

Prerequisites: Two-dimensional trajectory problems, velocity, acceleration, and differential equations. Read Sections 10.1, 10.3 and 710.4 in Stewart's **Calculus**.

Background: Imagine that a baseball player is up at home plate and hits the ball in the air. What parametric equations describe the position of the ball t seconds after it is hit? How far will the ball travel? How fast does the ball need to be hit in order for the ball to clear the home run fence? The following discussion and exercises are designed to answer these questions.

First consider a simplified model that ignores air resistance. In this case, after the ball is hit, the only force acting on the ball is the vertical force due to gravity. Therefore, the following equations describe the x and y components of the acceleration of the ball

$$\frac{d^2x}{dt^2} = 0 \quad \text{and} \quad \frac{d^2y}{dt^2} = -g$$

where g is the acceleration constant due to gravity ($g = 32$, where the unit of distance is feet). Integrating these equations with respect to t gives

$$\frac{dx}{dt} = C_1 \quad \text{and} \quad \frac{dy}{dt} = -gt + C_2$$

where the constants C_1 and C_2 can be evaluated by considering the initial velocity of the ball. Suppose the initial speed of the ball is the constant v (in units of feet per second) and suppose the angle of inclination of the ball is A. Then the x- and y- components of the initial velocity of the ball are given by $v \cos(A)$ and $v \sin(A)$, respectively. After substituting $t = 0$ in the above equations and solving for C_1 and C_2, the following equations are obtained

$$\frac{dx}{dt} = v \cos(A) \quad \text{and} \quad \frac{dy}{dt} = -gt + v \sin(A)$$

One more integration with respect to t yields

$$x = vt \cos(A) + c_1 \quad \text{and} \quad y = \frac{-gt^2}{2} + vt \sin(A) + c_2$$

Assume that the origin ($x = 0, y = 0$) is the location of home plate and that the shoulder height of the batter (from where the ball is hit) is h feet. Then by substituting $t = 0$, the constants c_1 and c_2 can be found. The final parameterization of the baseball is given by

$$x = vt \cos(A) \quad \text{and} \quad y = \frac{-gt^2}{2} + vt \sin(A) + h$$

Assignment: Solve the following problems.

1. Set $g = 32$, the angle A at $\pi/4$, the velocity v to be 120 feet per second, and the height h to be 6 feet. Define the x and y coordinates of the ball t seconds later (see the equation for x and y above) as functions of t, and then plot the trajectory of the ball until the ball hits level ground. Start the parameter t at 0, and experiment with different terminal values of t to try and get the entire flight of the baseball (until the ball hits the ground) on the screen. Find the horizontal distance traveled by the ball and the elapsed time.

2. What is the shape of the graph? Eliminate t, and determine the equation of the trajectory in the form of y as a function of x. To do this, solve for t in terms of x (this you can do by hand) and then substitute this expression for t into y.

   Now make the angle A and the initial speed v and t free variables and rewrite the equations for x and y.

3. Suppose the home run fence is 10 feet high and 350 feet from home plate. What is the minimum velocity at which the ball must leave the bat so that the ball barely clears the home run fence? *Be careful:* do not assume any particular value of the angle A. In fact your strategy should be as follows. First, eliminate t as you did in part 2 (only now using v and A as free variables). Then solve for the speed v in terms of the angle A by substituting $x = 350$, $y = 10$ and solving the resulting equation. Finally, minimize v as a function of A.

4. Now assume air resistance acts on the ball. Air resistance acts in the opposite direction to the velocity of the ball and its magnitude is proportional to the speed. This leads to the acceleration equations

$$\frac{d^2x}{dt^2} = -k\frac{dx}{dt} \quad \text{and} \quad \frac{d^2y}{dt^2} = -g - k\frac{dy}{dt}.$$

   Here, k is a friction constant, which will be given later. Solve these equations for x and y (using the initial conditions). Take the limit of your solutions as $k \to 0$ and see if your result agrees with the solution for x and y without air resistance. Then repeat parts 1 and 3, taking into account air resistance with $k = 0.1$. Compare your plots and your answers to your results without air resistance. For part 3, you will not be able to algebraically solve for v in terms of A. Instead, take specific values of A (near $\pi/4$) and solve for v numerically. Determine an approximate minimum value for v.

Related Activities: See the Applied Projects: *Which is Faster, Going Up or Coming Down?* (at the end of Section 10.3) and *Calculus and Baseball* (at the end of Section 10.4) in Stewart's **Calculus**.

10.18 Logistic Growth

Prerequisite: First order separable differential equations. Read Sections 10.4 and 10.5 in Stewart's **Calculus**.

Background: The logistic equation is a first order nonlinear differential equation of the form

$$\frac{dP}{dt} = aP - bP^2$$

which is often used to study population growth. The parameters a and b are positive constants that are related to the birth and death rate of the population. In the problem that follows, $P(t)$ represents the population of the United States at time t, in years.

The table below contains population data for the United States with population values given in millions. **Re-scale the time variable so that $t = 0$ corresponds to the year 1790.**

Year	Population	Year	Population
1790	3.93	1900	75.99
1800	5.3331	1910	91.97
1810	7.24	1920	105.71
1820	9.64	1930	122.78
1830	12.87	1940	131.67
1840	17.07	1950	151.333
1850	23.19	1960	179.32
1860	31.44	1970	203.21
1870	39.82	1980	226.5
1880	50.16	1990	249.63
1890	62.95	2000	?

Assignment: The goal of this project is to find positive constants a and b such that the solution to the logistic differential equation above satisfying $P(0) = 3.93$ *best fits* the data in the table above. One method for achieving a *best fit* is to proceed as follows. Begin by naming the population values in the table above

$$P_0, P_1, ..., P_{20}$$

That is, denote $P_0 = 3.93$, $P_1 = 5.3331$, etc. Now, find the solution $P(t)$ of the logistic equation above satisfying $P(0) = 3.93$ as a function of a, b and

t, and consider the function given by

$$F(a, b) = \sum_{n=1}^{20} \left(P(10n) - P_n \right)^2$$

Determine a mechanism for finding the positive values of a and b which minimize $F(a, b)$.

Use these values of a and b to estimate the population of the United States in the year 2000. Do you think this is a reasonable model for predicting the population in the year 2000?

10.19   Pension Funds

Prerequisites: Linear differential equations and exponential growth. Read Sections 10.4 and 10.6 in Stewart's **Calculus**.

Background: A pension fund starts out with $\$P$ (at $t = 0$) and is invested with a return of $100r\%$ per year, compounded continuously (here, r is the interest rate, given as a number between 0 and 1). The pension fund must continuously pay out money at the rate of $\$R$ per year to its employees for a period of n years and the value of the pension fund must decrease zero after n years. Let $y(t)$ denote the value of the pension fund after t years.

Assignment: Answer the following questions with complete sentences.

1. From the information given, derive the differential equation $y' = ry - R$, with the conditions $y(0) = P$, $y(n) = 0$.

2. Solve this differential equation for y.

3. Find a formula for P in terms of r, n, and R (P represents the amount of money required to pay out $\$R$ per year for n years, assuming the rate of return on the investment is $100r\%$).

4. Calculate P for $R = 50K$, $r = .07$, and $n = 20$.

5. Calculate the interest rate (r) required so that an initial value of $P = 500K$ for the pension fund will pay out $50K$ per year for 30 years.

10.20   Parachuting

Prerequisites: Max/min problems and first order, separable differential equations. See Sections 10.1 and 10.3 in Stewart's **Calculus**.

Background: A sky-diver, weighing 70kg., jumps from a plane at an altitude of 2000 feet and free falls for $T1$ seconds before pulling the rip chord. The parachute takes three seconds to open. A landing is defined to be "gentle" if the velocity on impact is less than the impact velocity of an object dropped (free-fall) from a height of 6 meters.

The distance traveled by the sky-diver t seconds after the jump can be found from Newton's law, *Force = Mass $\times$ Acceleration*. During the free-fall part of the jump, we will assume there is no air resistance. In this case, *Force* equals $-mg$, where g is the gravitational constant, 9.8 (meters/per second^2) and $m = 70$ kg. During the parachute phase of the jump, a significant "drag" term due to the air resistance of the parachute affects the *Force*. During this phase

$$Force = -mg - kv,$$

where v is the velocity and where k is the (positive) drag coefficient. Note that since velocity is negative (since the sky-diver is falling), the drag term, $-kv$, is positive (or up) thus tending to slow down the fall. During the three second period while the parachute is opening assume the drag coefficient is $k = 25$kg/sec. When the parachute is fully opened, assume $k = 110$kg/sec.

Assignment: Find the maximum time, $T1$, before pulling the rip chord so that the landing is gentle.

Hint: First find the maximum velocity for a gentle landing. Next, find the velocity after $T1$ seconds of free-fall. Use this velocity as the initial velocity for the differential equation that governs the three second period that the parachute is opening. The velocity at the end of this period can then be used as the initial velocity for the differential equation for the time period when the parachute is open. The velocity should be a function of $T1$ which you can then equate to the maximal gentle landing velocity on impact; and solve for $T1$.

10.21 Radioactive Waste at a Nuclear Power Plant

Prerequisites: First order linear differential equations. Read Section 10.4 in Stewart's **Calculus**.

Background: A nuclear power plant produces a waste product that is a radioactive isotope, called A. The isotope A has a half-life of 10 years; 70% (by weight) decays into a radioactive isotope B, and 30% decays into a radioactive isotope C. The isotope B has a half-life of 20 years and decays into nonradioactive by-products. The radioactive isotope C has a half-life of 30 years and also decays into nonradioactive by-products.

Assume that isotopes B and C weigh essentially the same as isotope A. Thus, for example, if 100 kilograms of A decays there will be 70 kilograms of B and 30 kilograms of C.

Assignment: Answer the following questions.

1. Suppose you start with 400 kilograms of isotope A. What is the maximum amount of isotope B that will be present, and when will this occur? Answer the same question for isotope C.

2. When the power plant was first turned on, there was no isotope A, B, or C present. If the power plant operates so that it produces isotope A at the constant rate of 40 kilograms per year, what are the maximum amounts of isotopes A, B, and C that will be present, and when will these different maxima occur?

3. Federal safety requirements say that the reactor can never have on hand more than 500 kilograms of isotope A, 400 kilograms of isotope B, or 300 kilograms of isotope C. What is the maximum rate at which the power plant can produce isotope A without violating the federal regulations?

Hints:

(a) First find the three decay constants.

(b) If a radioactive isotope is being produced by some source at the same time as it is decaying, how does that alter the differential equation for the rate of change of the amount of this isotope?

(c) Be sure to plot A, B, and C, in order to ascertain if they have the qualitative behavior you expect.

10.22   Estimating Infinite Sums

Prerequisites: Read about infinite sums in Chapter 8 of this manual and Chapter 12 in Stewart's **Calculus.**

Background: A calculator or a computer can be used to approximate the infinite sum $\sum_0^\infty a_k$ with the finite sum $\sum_0^N a_k$. The number N is chosen so that the remainder term, $\sum_{N+1}^\infty a_k$, is smaller than some given tolerance.

Assignment:

1. The same principals used in showing how the integral test works can be used in choosing N for approximating the sum. Draw a picture using rectangles, much like that done in studying the integral test, to show that

$$\sum_{k=N+1}^\infty a_k < \int_N^\infty f(x)\,dx$$

   where $a_k = f(k)$, and we assume that $f(x)$ is a positive decreasing function. This shows that the terms in the infinite sum that are neglected are smaller than the integral. In addition, this shows that the integral is a measure of the error in this approximation.

2. Use any the integral test to show that the following series converge:

$$\sum_{k=1}^\infty \frac{1}{k^3} \qquad \sum_{k=1}^\infty ke^{-k^2} \qquad \sum_{k=2}^\infty \frac{1}{k(\ln k)^4}$$

3. Use the inequality in question 1 and your calculator to approximately evaluate the sums in question 2 to three decimals of accuracy. To do this, evaluate the integral for arbitrary N and then choose N so that the integral is less than 0.5×10^{-3}.

4. Modify the solution to question 1 to show that

$$\int_{N+1}^\infty f(x)\,dx < \sum_{k=N+1}^\infty a_k < \int_N^\infty f(x)\,dx$$

5. Use the inequality above to argue that

$$\sum_{k=1}^{N} a_k + \frac{1}{2}\left(\int_{N}^{\infty} f(x)\,dx + \int_{N+1}^{\infty} f(x)\,dx \right)$$

approximates $\sum_{k=1}^{N} a_k$ with an error which is no greater than

$$\frac{1}{2} \int_{N}^{N+1} f(x)\,dx$$

6. Use the estimate and error term in question 5 to find a values of N so that the estimate approximates the series in question 2 within 0.5×10^{-3}.

10.23   Power Series and Convergence

Prerequisites: Read about infinite series in Chapter 8 of this manual and Chpater 12 of Stewart's **Calculus**.

Background: Taylor polynomials can be used to approximate a smooth function. The corresponding Taylor series only will agree with the function for those values of x for which the series converges. The degree of the approximating polynomial is very dependent on the value of x under consideration.

Assignment:

1. Compute the general formula for the coefficients of the Taylor series for the function
$$f(x) = \frac{x}{x+1}$$
centered at 0. Be sure to have this part correct as the other questions depend on it!

2. What degree Taylor polynomial (centered at 0) is needed to approximate this function correct to three decimal places at a) $x = 0.1$, b) $x = 0.3$, c) $x = 0.5$ and d) $x = 0.8$? Plot graphs of $f(x)$ and the polynomial approximation for each case. Comment on any trend you see in parts a) through d).

3. Find the greatest value of x for which a) $P_{25}(x)$ b) $P_{40}(x)$ approximates the correct function values of $f(x)$ with an accuracy of at least three decimal places. Plot the error $|P_n(x) - f(x)|$ for each approximation. Comment on how the degree of the polynomial affects the accuracy.

4. Does it make sense to ask the question: What degree Taylor polynomial centered at 0 is needed to approximate a) $f(1)$ b) $f(1.5)$ correct to three decimal places? Explain.

Index